SpringerBriefs in Molecular Science

For further volumes:
http://www.springer.com/series/8898

Rui-Qin Zhang

Growth Mechanisms and Novel Properties of Silicon Nanostructures from Quantum-Mechanical Calculations

 Springer

Rui-Qin Zhang
Department of Physics and
 Materials Science
City University of Hong Kong
Hong Kong SAR
People's Republic of China

ISSN 2191-5407   ISSN 2191-5415 (electronic)
ISBN 978-3-642-40904-2 ISBN 978-3-642-40905-9 (eBook)
DOI 10.1007/978-3-642-40905-9
Springer Heidelberg New York Dordrecht London

Library of Congress Control Number: 2013953595

Printed on acid-free paper

Springer is part of Springer Science+Business Media (www.springer.com)

Foreword

The shrinkage of dimensions of nanomaterials is expected to reveal new fascinating properties which may ultimately lead to exciting applications in optoelectronic, nanoelectronic, environmental, energy, biological, and medical areas. Different aspects of nanoscience and nanotechnology (nanomaterials structure, growth mechanisms, and properties) can now be studied by modern computational methods, thereby providing new insights at the atomic level which are otherwise unavailable or difficult to obtain.

In this SpringerBriefs, entitled "Growth Mechanisms and Novel Properties of Silicon Nanostructures from Quantum-Mechanical Calculations," Prof. R. Q. Zhang provides a comprehensive review of theoretical studies on this particular topic performed by his group over the years. The book is divided into Five chapters: namely Chap. 1, "Introduction," Chap. 2, "Growth Mechanism of Silicon Nanowires," Chap. 3, "Stability of Silicon Nanostructures," Chap. 4, "Novel Electronic Properties of Silicon Nanostructures," Chap. 5, "Summary and Remarks." The specific subject matters covered are: the growth mechanisms and properties of Si quantum dots, nanotubes and nanowires, including mechanisms of oxide-assisted growth of silicon nanowires, thin stable short silicon nanowires, energetic stability of silicon nanotubes, thermal stability of hydrogen terminated silicon nanostructures, size-dependent oxidation of silicon nanostructures, excited-state relaxation of hydrogen terminated silicon nanodots, and direct-indirect energy band transitions of silicon nanowires by surface engineering and straining in silicon nanostructures. In this regard, this SpringerBriefs represents an authoritative, systematic, and detailed description of the subject matters.

The book is highly valuable to scientists and graduate students working in such diverse fields as nanoscience and nanotechnology, surface science and technology, semiconductor technology, nanomaterials and nanofabrication, computational physics and chemistry, etc.

Prof. Boon K. Teo

Contents

Chapter 1
Introduction

Abstract It was the quantum confinement effect and photoluminescence that inspired the intensive researches in the research field of growth and properties of silicon nanostructures. A large volume of researches have been directed to experimental syntheses and characterizations of the various silicon nanostructures including zero-dimensional quantum dots and one-dimensional nanowires. Computational predictions of novel structures of pristine silicon nanostructures including silicon nanotubes have also been intensive. Distinguishing computational work has been done by us on the growth mechanism, surface properties, excited state properties, and energy band engineering of silicon nanostructures. Our studies are expected to promote the development of silicon-based nanoscience and nanotechnology.

Keywords Silicon quantum dots · Silicon nanowires · Quantum confinement · Photoluminescence · Growth mechanism · Surface · Excited state properties · Energy band engineering

The shrinkage of dimensions of nanomaterials is expected to result in new fascinating properties related to size and quantum effects, which may lead to exciting applications in optoelectronics, chemical and biological sensing, etc. In recent decades, silicon nanomaterials and nanostructures have attracted extensive interest, not only for the technological development, but also for the researches on fundamental sciences. Manifold interesting changes show up as the silicon-based materials shrink down to the nanoscale. Consequently, low-dimensional nanomaterials, such as zero-dimensional (0D) quantum dots (SiQDs) [1], one-dimensional (1D) silicon nanowires (SiNWs) [2, 3], and two-dimensional (2D) silicon sheets [4], exhibit dimension-tunable electronic and optical properties, which are important attributes in nanodevice applications.

The main reasons that made the study of silicon nanostructures an active field of research long time ago are the strong room-temperature photoluminescence (PL) and the observation of quantum size effect, for example, in silicon cluster assembled films. Due to the easier control in size of the deposited silicon clusters than that of porous silicon, silicon cluster films allow a more direct study of the

R.-Q. Zhang, *Growth Mechanisms and Novel Properties of Silicon Nanostructures from Quantum-Mechanical Calculations*, SpringerBriefs in Molecular Science, DOI: 10.1007/978-3-642-40905-9_1, © The Author(s) 2014 quantum confinement effect. Films composed of clusters larger than 3 nm presented PL following the quantum confinement model closely [5–7]; However, this is no longer the case for films containing smaller clusters [8, 9], probably due to the deviation of the smaller cluster's (<2.0 nm, 200 atoms) structures from a diamond structure, as evidenced in transmission electron microscopy (TEM) studies [9, 10].

Later, silicon nanoparticles less than 4 nm in diameter were shown to present excellent luminescence properties in the visible region [11, 12]. SiNWs of different growth directions and diameters can have large and direct bandgaps. We have also revealed interesting size-dependent optical properties of SiQDs [13, 14] and SiNWs [15]. The absorption energies of SiQDs showed a blueshift with decreasing particle size and different degree of localization due to excite state relaxation. The formation of self-trapped exaction is likewise observed for relatively shorter SiNWs, which may have potential application in efficient exciton transport as a result of strong quantum confinement. The ability to manipulate quantum effects is considered to be critical for technologies that will form the basis for the next generation of computing, optical, and electronic devices.

Although SiNWs could be the key components of the next generation electronics, there are still a number of challenges being encountered by the conventional silicon-based electronic technology to have the feature size going down to tens of nanometer or even smaller in CMOS logic circuits. It is expected that the device working principles, physical limitation in production, and increased production cost would all be concerns. Nanotechnology is seen to play an important role in pushing the functional density and data transfer rate to a higher level. Many new phenomena in electrical, optical, mechanical, and chemical properties will emerge in the nano regime, in addition to the already revealed quantum confinement effect and size effect [16], giving hope to future nanodevices to be of higher speed, low power consumption, higher efficiency, higher integration, and more economically attractive.

To reach the goal, developments of controlled synthesis/fabrication of nanomaterials are critical. Various kinds of silicon nanostructures have been obtained, including SiQDs [12] and semiconductor nanowires [2, 17]. The used synthesis techniques include photolithography [18] and scanning tunneling microscopy (STM) [19] which, however, can only produce nanomaterials in low quantity, not suitable for industrial application.

The growth mechanism of one-dimensional (1D) nano materials (such as: CNT) usually follows a vapor–liquid–solid (VLS) model, proposed in the 1960s [20], which requires the use of metals catalysis. There have also been a screw dislocation-assisted growth mechanism [21] following which one can grow 1D nanomaterials of large diameter and with whiskers or rod morphology.

The morphology and structure of the synthesized nanowires can be systematically characterized using scanning electron microscope (SEM) and TEM. There have been SiNWs synthesized presenting single Si core crystal/oxide shell structures and containing no metal particles in the wire tip, with smooth surface and almost similar diameters. Their growth direction is <112> and the tips of SiNWs have high density of defects, showing that the defects may play a role in assisting the growth of SiNWs [22]. Observed also include SiNWs looking like spring, fish bones, fog eggs, and chains. Using photoluminescence spectra, the SiNWs were confirmed to consist of a silicon crystal core and a covered oxide layer. X-ray absorption spectra further indicated that the silicon crystal core is wide band-gapped, due to the decreased crystallinity [23]. Replacing the outer oxide layer by a crystalline β-SiC thin layer can increase and stabilize the PL signals [24].

Contrary to the abundant experimental work of silicon nanostructures, theoretical efforts devoted to revealing the growth and physical properties are rather limited and mostly on pristine SiNWs. Menon and Richter have reported quasi 1D Si structures by using a generalized tight binding molecular dynamics scheme [25]. The proposed Si nanowire geometries of surfaces closely resemble one of the most stable reconstructions of the crystalline Si surfaces with a core of buck-like fourfold coordinated atoms. Marsen and Sattler [26] predicted that the wires of 3–7 nm in diameter and at least 100-nm-long tend to be assembled in parallel in bundles, and they proposed a fullerene-type Si_{24}-based atomic configuration for the nanowires. We also studied the geometric structures of thin short silicon nanowires consisting of tricapped trigonal prism Si_9 subunits and uncapped trigonal prisms [27]. These structures are the thinnest stable silicon nanowires, which are much more stable than the silicon nanotubes built analogously to small carbon nanotubes. Zhao and Yakobson [28] compared the energetics of the formation of SiNWs with crystalline and polycrystalline cores and suggested that for very thin (1–3 nm) nanowires, polycrystalline structure is lower in energy than the single crystal structure observed in experiments. Later, they reported that a polycrystalline silicon wire of fivefold symmetry, rather than single-crystal types, represents the ground state for diameter less than 6 nm [29]. Menon et al. [30] compared the tetragonal and cage-like or clathrate nanowires of Si and suggested that cage-like nanowires possessing lower density can maintain their structural integrity over a larger range of strain conditions than the tetrahedral nanowires. Ponomareva et al. reported the polycrystalline forms of nanowires, with the smallest surface to bulk ratio are also the least stable, while the tetrahedral type nanowires are found to be the most stable [31]. Recently, Ponomareva reported that the tetrahedral type nanowires oriented in the <111> direction were the most stable, and the stability of the cage-like nanowires lie somewhere between this and tetrahedral nanowires oriented in other directions [32]. Kagimura and co-workers reported that for the several structures of pristine Si and Ge nanowires with diameters between 0.5 and 2.0 nm diamond-structured nanowires are unstable for diameters smaller than 1 nm, filled-fullerene wires are the most stable for diameters between 0.8 and 1 nm, and a simple hexagonal structure is particularly stable for even smaller diameter [33]. Nishio et al. proposed a novel polyicosahedral nanowire which is energetically advantageous over the pentagonal one for a wire whose diameter is less than 6.02 nm based on a series of annealing molecular dynamics simulations [34]. Cao et al. performed extensive *ab initio* studies to search for stable geometries of <100> SiNWs and reported that Si nanowires with diameters smaller than 1.7 nm prefer to adopt a shape with a square cross-section

[35]. Fthenakis and co-workers found 17 optimum Si_{38} fullerene isomers constructed with permutations among their pentagons and hexagons [36] using a global optimization method and the isomer with three distinct fused quadruples of pentagons is energetically optimal.

Among the theoretical researches on structural properties of hydrogenated silicon systems, detailed structures and energetics of very small Si_nH_x (n < 16) clusters have been carried out by the author and others using quantum chemical calculations [37–40]. In particular, we showed that in order to maintain a stable nanometric and tetrahedral silicon crystallite and remove the gap states, the saturation atom or species such as H, F, Cl, OH, O, or N is necessary. Klein, Urbassek, and Frauenheim systematically studied the structural and dynamic properties of a-Si:H using a tight-binding Hamiltonian constructed based on calculations with the density functional approach [41, and references therein] and found that Si atoms are mainly fourfold coordinated, with around 7 % fivefold coordinated atoms, while H atoms exhibit a tendency for clustering. Kratzer et al. studied the reaction dynamics of atomic hydrogen with the hydrogenated Si(001) surfaces [42], and the H_2 adsorption and desorption on Si(001) surfaces [43]. Lee et al. studied the role of hydrogen using first-principles theory for Si adatom adsorption and diffusion on hydrogenated Si(001) surfaces [44, 45].

Regarding theoretical studies on electronic and optical properties, attempts were made with various theoretical methods including generalized gradient approximation (GGA) of density functional theory (DFT), diffusion quantum Monte Carlo (QMC) technique, GW correction, and time-dependent local density approximation (TDLDA), etc. [46–48]. The bandgaps of silicon nanostructures were found blue-shifted from the infrared to the visible region as the size was reduced, consistent with experiments. However, numerous controversies still exist between theoretical and experimental results, mainly due to: (i) the uncertain particle size in the measurement, (ii) the surface impurity of the hydrogenated silicon particles, and (iii) the approximations of the contemporary approaches for studying the excited state properties. Puzder et al. [49] computed the absorption and emission energies of small silicon nanocrystals, using LDA and QMC methods, which demonstrated that the optical characters especially their absorption and emission energies are sensitive to the size, surface structure, and chemistry of the semiconductor nanoparticles. It has been widely recognized that the difference of excited states determined in different calculations is mainly due to the different approximations adopted for spectrum calculations that include the treatment of Coulomb or excitonic interactions. With the increased computational power, the large size silicon nanoparticles beyond those with diameters less than 1 nm (such as $Si_{29}H_{24}$, $Si_{29}H_{36}$, and $Si_{35}H_{36}$) are being paid increasing attention. Chelikowsky and his co-workers [12] examined the TDDFT, in particular, the TDLDA, for the optical properties of silicon nanoparticles with diameters of $0 \sim 2$ nm. It was considered that the TDLDA method that utilizes a real-space description of the electronic structure problem is better than QMC in offering an efficient approach for large systems. Although silicon nanoparticles have become a subject of many computational studies at ab initio and DFT levels, the origin and the physical or chemical mechanisms of their outstanding optical properties still require considerable efforts.

Over the years, we have performed systematic computational studies on different aspects of silicon nanoscience and technology (nanomaterial structures, growth mechanisms, and properties), aiming at filling the deficiencies in the area. Our results can provide insight and predict properties critically needed by experimentalists. In this Brief, we review our computational studies that address different aspects of science and technology of SiQDs and SiNWs. The studies include mechanisms of oxide-assisted growth of silicon nanowires, thin stable short silicon nanowires, energetic stability of silicon nanotubes, effects of structural saturation of silicon nanostructures, thermal stability of hydrogen-terminated silicon nanostructures, size-dependent oxidation of silicon nanostructures, excited state relaxation of hydrogen-terminated silicon nanodots, and direct–indirect energy band transitions of silicon nanowires by surface engineering and straining. We also discuss their potential applications.

References

1. Wolkin MV, Jorne J, Fauchet PM, Allan G, Delerue C (1999) Phys Rev Lett 82:197
2. Morales AM, Lieber CM (1998) Science 279:208
3. Zhang YF, Tang YH, Wang N, Yu DP, Lee CS, Bello I, Lee ST (1998) Appl Phys Lett 72:1835
4. Kara A, Léandri C, Dávila M, De Padova P, Ealet B, Oughaddou H, Aufray B, Le Lay G (2009) J Supercond Nov Magn 22:259
5. Ehbrecht M, Kohn B, Huisken F, Laguna MA, Paillard V (1997) Phys Rev B 56:6958
6. Ehbrecht M, Huisken F (1999) Phys Rev B 59:2975
7. Ledoux G, Guillois O, Porterat D, Reynaud C, Huisken F, Kohn B, Paillard V (2000) Phys Rev B 62:15942
8. Mélinon P, Kéghélian P, Prével B, Perez A, Guiraud G, LeBrusq J, Lermé J, Pellarin M, Broyer M (1997) J Chem Phys 107:10278
9. Mélinon P, Kéghélian P, Prével B, Dupuis V, Perez A, Champagnon B, Guyot Y, Pellarin M, Lermé J, Broyer M, Rousset JL, Delichère P (1998) J Chem Phys 108:4607
10. Goldstein AN (1996) Appl Phys A Mater Sci Process 62:33
11. Chelikowsky JR, Kronik L, Vasiliev I (2003) J Phys: Condens Matter 15:R1517
12. Gupta Anoop, Swihart Mark T, Wiggers Hartmut (2009) Adv Funct Mater 19:696–703
13. Wang X, Zhang RQ, Lee ST, Niehaus TA, Frauenheim T (2007) Appl Phys Lett 90:123116
14. Wang X, Zhang RQ, Niehaus TA, Frauenheim T, Lee ST (2007) J Phys Chem C 111:12588
15. Wang Y, Zhang RQ, Frauenheim T, Niehaus TA (2009) J Chem Phys C 113:12935
16. Waser R (2002) Nanoelectronics and information technology: materials, processes, devices. Wiley, Weinheim
17. Zhang YF, Tang YH, Wang N, Yu DP, Lee CS, Bello I, Lee ST (1998) Appl Phys Lett 72:1835
18. Liu HI, Maluf NI, Pease RFW (1992) J Vac Sci Technol B10:2846
19. Ono T, Saitoh H, Esashi M (1997) Appl Phys Lett 70:1852
20. Wagner RS, Ellis WC (1964) Appl Phys Lett 4:89
21. Frank FC (1949) Discov Faraday Soc 5:48
22. Wang N, Zhang YF, Tang YH, Lee CS, Lee ST (1998) Phys Rev B58:R16024

23. Zhang YF, Liao LS, Chan WH, Lee ST, Sammynaiken R, Sham TK (2000) Phys Rev B 61:8298
24. Zhou XT, Zhang RQ, Peng HY, Shang NG, Wang N, Bello I, Lee CS, Lee ST (2000) Chem Phys Lett 332:215–218
25. Menon M, Richter E (1999) Phys Rev Lett 83:792
26. Marsen B, Sattler K (1999) Phys Rev B 60:11593
27. Li BX, Cao PL, Zhang RQ, Lee ST (2002) Phys Rev B 65:125305
28. Zhao Y, Yakobson BI (2003) Phys Rev Lett 91:035501
29. Zhao Y, Yakobson BI (2005) Phys Rev Lett 95:115502
30. Menon M, Srivastava D, Ponomareva I, Chernozatonskii LA (2004) Phys Rev B 70:125313
31. Ponomareva I, Menon M, Srivastava D, Andriotis AN (2005) Phys Rev Lett 95:265502
32. Ponomareva I, Menon M, Richter E, Andriotis AN (2006) Phys Rev B 74:125311
33. Kagimura R, Nunes RW, Chacham H (2005) Phys Rev Lett 95:115502
34. Nishio K, Morishita T, Shinoda W, Mikami M (2006) J Chem Phys 125:074712
35. Cao JX, Gong XG, Zhong JX, Wu RQ (2006) Phys Rev Lett 97:136105
36. Fthenakis ZG, Havenith RW, Menon M, Fowler PW (2007) Phys Rev B 75:155435
37. Zhang RQ, Costa J, Bertran E (1996) Phys Rev B 53:7847
38. Onida G, Andreoni W (1995) Chem Phys Lett 243:183
39. Miyazaki T, Uda T, S˘tich I, Terakura K (1996) Chem Phys Lett 261:346
40. Meleshko V, Morokov Yu, Schweigert V (1999) Chem Phys Lett 300:118
41. Klein P, Urbassek HM, Frauenheim Th (1999) Phys Rev B 60:5478
42. Kratzer P (1997) J Chem Phys 106:6752
43. Kratzer P, Hammer B, Nørskov JK (1995) Phys Rev B 51:13432
44. Lee SM, Lee YH, Kim NG (2000) Surf Sci 470:89
45. Kratzer P, Pehlke E, Scheffler M, Raschke MB, HWfer U (1998) Phys Rev Lett 81:5596
46. Lehtonen O, Sundholm D (2005) Phys Rev B 72:085424
47. Williamson AJ, Grossman JC, Hood RQ, Puzder A, Galli G (2002) Phys Rev Lett 89:196803
48. Degoli E, Cantele G, Luppi E, Magri R, Ninno D, Bisi O, Ossicini S (2004) Phys Rev B 69:155411
49. Puzder A, Williamson AJ, Grossman JC, Galli G (2003) J Am Chem Soc 125:2786

Chapter 2
Growth Mechanism of Silicon Nanowires

Abstract Among the variety of nanostructural synthesis methods, oxide-assisted growth is unique in terms of not only the quality and quantity of nanostructures synthesized but also its interesting and novel mechanism. It has been revealed by our theoretical calculations that it is the silicon suboxide clusters which possess high reactivity on their surface silicon sites that facilitate the nucleation and growth of silicon nanostructures by allowing them to grow and form sp^3 cores after a critical size. The high possibility and crystallographic dependence of oxygen diffusion allow the so-formed silicon nanostructures grow along certain growth directions (<110> and <112>).

Keywords Oxide-assisted growth · Silicon oxide clusters · Oxygen diffusion · Nucleation · Reactivity

Large quantity of SiNWs have been synthesized more than 10 years ago by laser ablation using metal containing semiconductor targets [1, 2], following the traditional VLS growth mechanism in which metal or metal compounds are the catalysts for the nanowires growth process [3]. Soon afterward, it was found that the nanowires could be synthesized without any metal [4, 5]. The oxide and crystalline defects were found to play an important role in the nucleation and growth process. The so-synthesized nanowires are not only in large quantity (10 times of the metals catalysis synthesis technique) but also in high quality without any metal impurities. The related research in nucleation of nanomaterials and their growth mechanism has become important topics of the field.

The oxide-assisted growth mechanism has the following unique characteristics over the VLS growth mechanism: (1) reduced metal contamination due to the exclusion of using metal catalyst; (2) rich source materials due to the use of oxide; and (3) capability to produce SiNWs in <112> and<110> directions in contrast to the VLS growth which is mainly in <111> direction.

Obviously, during the synthesis of Si nanowire growth (1) oxygen atoms desorbed from SiO_2 during laser ablation at high temperature may contribute to the growth; (2) different species of Si oxides may have been formed in the presence of oxygen; (3) SiO_2 may change the ability of the target to absorb energy from the

R.-Q. Zhang, *Growth Mechanisms and Novel Properties of Silicon Nanostructures*
from Quantum-Mechanical Calculations, SpringerBriefs in Molecular Science,
DOI: 10.1007/978-3-642-40905-9_2, © The Author(s) 2014

laser beam; and (4) reaction of Si and SiO_2 to form oxides in vapor phase. Thus, the influence of oxide on the formation and growth of Si nanowires is critical here.

Silicon oxide clusters generated and present in the gas phase in Si nanowire synthesis play an important role in the nucleation and growth. There have been rich literatures studying small silicon oxide clusters Si_nO_m (n, m = 1–8) both experimentally and theoretically [6–13]. In particular, our systematic theoretical research [13] has revealed that silicon monoxide like clusters adopt planar and buckled-ring configurations, while oxygen-rich clusters are rhombuses arranged in a chain with adjacent ones perpendicular to each other.

Based on the calculation results on the gas-phase silicon oxide clusters, we revealed the underneath mechanism at an atomic level. We obtained interesting results such as the gas-phase favorable composition and distinctive features in reactivity of the different silicon oxide clusters. Remarkably, we found that Si suboxide clusters are highly reactive to bond with other clusters and prefer to form Si–Si bonds [14]. Specifically, the analysis of the highest occupied molecular orbitals (HOMOs) and the lowest unoccupied molecular orbitals (LUMOs) of silicon oxide clusters uncovered the reactivity for them to form the Si–Si, Si–O, and O–O bonds, according to the well-known frontier orbital theory [15]. The much smaller HOMO-LUMO gap for $(SiO)_n$ clusters (2.0–4.5 eV) than those for $(SiO)_2$ species indicates higher chemical reactivity of the $(SiO)_n$ clusters. The localization of HOMO mainly on the Si atoms at the cluster surface makes these regions highly reactive. The silicon oxide cluster with the O ratio less than about 0.62 presents remarkably larger reactivity to form a Si–Si bond of two silicon oxide clusters than to form a Si–O or O–O bond [14], as shown in Fig. 2.1, facilitating the combination of these clusters through the Si–Si bonding.

Our result revealed that the cluster with a higher ratio of Si atoms has a higher chance to form a Si–Si bond with others. However, in the actual synthesis, the presence of the cluster is also determined by the energetic favorability. Thus, the higher cohesion energy per atom of the silicon-rich clusters indicates a smaller

Fig. 2.1 The inverse of the energy difference ΔE = LUMO (electron acceptor)-HOMO (electron donor) and thus the reactivity (proportional to the inverse of the energy difference) for the formation of a Si–Si bond, a Si–O bond, or an O–O bond between two silicon oxide clusters as a function of the Si:O ratio. Reprinted with permission from Ref. [14]. Copyright 2001, The American Physical Society

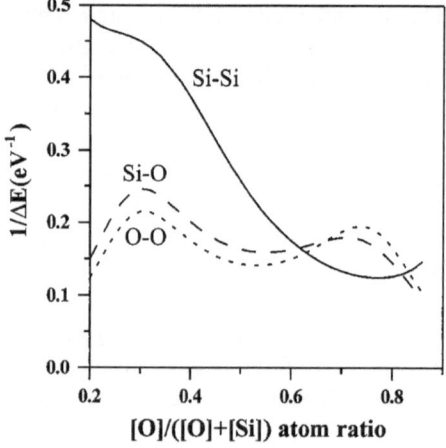

chance of their presence in the gas phase. The most probable cluster to achieve the highest yield and formation of Si nanowire should have an optimum ratio of Si atom to O atom in the silicon suboxide clusters close to 1, as also observed experimentally (about 49 at. % of O) [4, 5]. It is worthwhile noting that there are also experimental reports on the formation of the crystalline phase of Si nanoclusters from the deposition of silicon-rich oxide [16].

According to the above findings, we have elucidated the mechanism of oxide-assisted nucleation of silicon nanostructures according to our calculations. Briefly, it is the silicon suboxide cluster which possesses unsaturated feature and high reactivity on silicon atoms [14] that facilitates the formation of Si–Si bond with the other silicon oxide clusters. In contrast, oxygen-rich silicon oxide cluster prefer to form Si–O bond with the others. Accordingly, we have proposed a clear mechanism of silicon nanowire nucleation [9], in which some highly reactive silicon atoms in the silicon suboxide cluster deposited on the substrate would form bonds with the substrate atoms, sticking to the substrate and protecting the other reactive silicon atoms in the cluster from being deactivated. The latter silicon atoms face outside the substrate and thus can interact with other species in the vapor, acting as the nuclei and facilitating the formation of the silicon nanowires with a certain crystalline orientation.

The nucleation of Si nanocrystals could be expected to take place via the combination of small Si suboxide clusters. It has been further revealed that tetrahedral Si core begins to form at n = 5, as shown in Fig. 2.2 [17]. It is seen that (1) a Si core (represented by the open circles containing stars) is formed and is surrounded by a silicon oxide sheath; (2) the Si–Si bonds is formed mainly in the center to reduce the strain energy; (3) most of the Si atoms in the Si core have a tetrahedral coordination, similar to that in silicon crystal, quite different from that of pure Si clusters of the same size [18]; (4) the larger the size of the cluster is, the larger the size of the Si core increased and the fraction of Si atoms with three and four coordinates; and (5) beginning at n = 18, the sp^3 Si cores similar to the configuration in the Si crystal is formed, with all of the Si atoms in Si cores being four-coordinated. Figure 2.3 is the binding energies of $(SiO)_n$ clusters containing Si cores as a function of n. Those containing buckled structures are also provided. We found that: (1) the ones containing Si cores become energetically more favorable than the buckled ones for n = 5 and larger; and (2) the cluster becomes increasingly more stable with increasing Si core size. The two kinds of structures from n = 5 to n = 8 in Fig. 2.3 are close in energy and may compete in experiments. We obtained their relative population at 900 °C (the growth temperature of Si nanowires [19]) at equilibrium described by the Boltzmann factor $\exp(-E/kT)$, where k is Boltzmann's constant, E is the energy difference, and T is temperature in Kelvin. We confirmed that the structures containing Si cores still dominate at such a high temperature starting at a size as small as n = 8 (see the inset in Fig. 2.3). The formation of sp^3 Si core inside the silicon oxide clusters is important, as it contributes to the nucleation of the Si nanocrystals. Their high chemical reactivity allows these clusters to combine easily and to form a large sp^3 Si core via subsequent reconstruction and O migration from the center to the

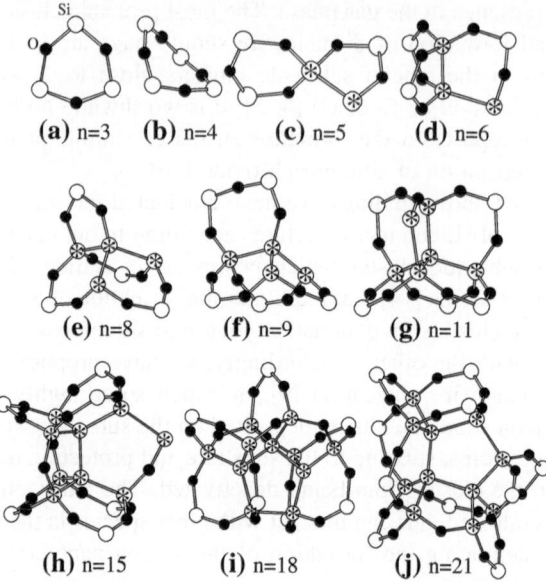

Fig. 2.2 The most favorable structures of silicon monoxide clusters $(SiO)_n$ for $n = 3–21$. Reprinted with permission from Ref. [17]. Copyright 2004, The American Physical Society

surface of the clusters. The crystalline Si cores thus formed can act as nuclei and precursors for subsequent growth of Si nanostructures.

We further studied three different isomers of the $(SiO)_{21}$ cluster with an O atom locating at different sites from the center to the surface of the cluster, which are shown in Fig. 2.4. We found that the most stable configuration is the one with O located on its surface, with a total binding energy of 211.74 eV. However, the binding energy decreases as the O atom moves from the surface into the cluster.

Fig. 2.3 Binding energy (eV/atom) of $(SiO)_n$ clusters versus n. The up triangles are $(SiO)_n$ with the Si-cored structure surrounded by a silicon oxide sheath, and open circles are those with buckled-ring structure. The inset shows the relative population of the former (N_Δ) and the latter (N_O) structures at 900 °C. Reprinted with permission from Ref. [17]. Copyright 2004, The American Physical Society

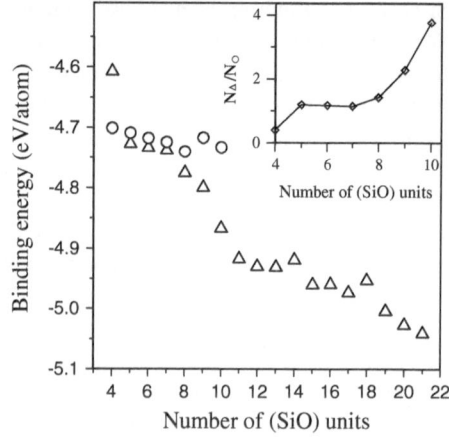

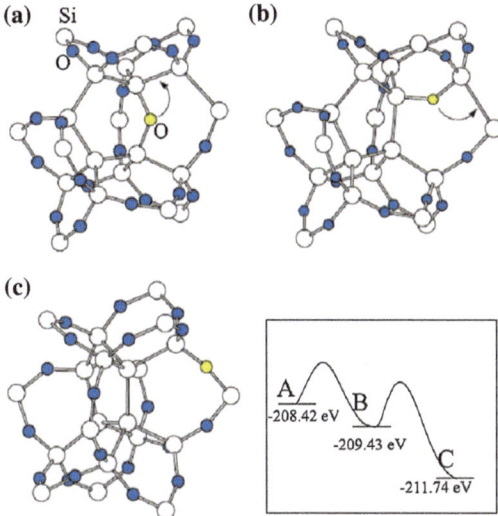

Fig. 2.4 Possible path of O atom migration from the center of a $(SiO)_n$ cluster to its surface: **a** $(SiO)_5$ and **b** $(SiO)_{21}$. Reprinted with permission from Ref. [17]. Copyright 2004, The American Physical Society

The result indicates that the O atom could migrate from the center of the silicon monoxide cluster to its surface via bond switching. The estimated migration barrier is about 1.79 eV for the $(SiO)_5$ cluster. The migration of O atom from the inside to the surface may be caused by the high strain involved in the large $(SiO)_n$ cluster, giving rise to the formation of a Si core. The Si core in the nuclei would grow larger with the assistance of O diffusion from the core to the surface layer during deposition.

With the above findings, we can describe what could happen in the synthesis of SiNWs using SiO powder or a mixture of Si and SiO_2 powder as the source. In the experiment, the evaporated $(SiO)_n$ clusters deposited on a substrate would be anchored due to their high reactivity at Si sites. The deposited clusters would act as the nuclei to absorb $(SiO)_n$ clusters from the vapor because of their remaining reactive Si atoms facing outward from the substrate. A Si core would start to form at a size of $n \sim 5$. The nuclei containing a Si core would grow larger with the assistance of O diffusion from the core to the surface layer during deposition. The O diffusion length depends on the temperature and the crystallographic orientation of the crystalline core formed, leading to the formation of Si nanowires with different crystalline orientations such as <110> and <112>, as observed in our experiments [19]. The above process may be similarly responsible for the ready formation of Si nanocrystals in the sp^3 configuration from amorphous SiO [20].

The catalytic effect of the Si_xO $(x > 1)$ layers on the nanowire tips is an important driving force for the nanowire growth. Among the number of different forms of Si suboxides, some of them are very reactive. The materials at the Si nanowire tips (similar to the case of nanoparticles) may be in or near their molten states. This is because that the surface melting temperatures of nanoparticles can be much lower than that of their bulk materials. For example, the difference between the melting temperatures of Au nanoparticles (2 nm) and Au bulk material is over 400 °C [21]. The atomic absorption, diffusion, and reaction are thus largely enhanced at the tips.

After nucleation, the further growth of the silicon domain may be crystallographic-dependent. The oxygen atoms in the silicon suboxide clusters during the growth of silicon nanowires might be expelled by the silicon atoms and diffuse to the edge in some directions of nanowires, forming the silicon oxide sheath. In a certain orientation, e.g., [112], the diffusion might be lower and the high reactive silicon oxide phase can still expose to the outside and facilitate the continuous growth of the wire in such a direction. However, the oxygen-rich sheath resulted in the other directions may possess lower reactivity and thus does not favor further stacking of silicon oxide clusters from the gas phase, leading to the growth suppression in such directions. The reactivity of silicon atoms in oxygen-rich clusters becomes very low at Si:O being 1:2. Instead, the reactivity of oxygen atoms increases a little. However, the overall reactivity is still quite low. It explains the retard of the shell silicon oxide on the sideways or lateral growth of nanowires.

References

1. Morales AM, Lieber CM (1998) Science 279:208
2. Zhang YF, Tang YH, Wang N, Yu DP, Lee CS, Bello I, Lee ST (1998) Appl Phys Lett 72:1835
3. Wagner RS, Ellis WC (1964) Appl Phys Lett 4:89
4. Wang N, Zhang YF, Tang YH, Lee CS, Lee ST (1998) Phys Rev B58:R16024
5. Zhang RQ, Lifshitz Y, Lee ST (2003) Adv Mater 15:635
6. Rao BK, Jena P (1985) Phys Rev B 32:2058
7. Ogut S, Chelikowsky JR, Louie SG (1997) Phys Rev Lett 79:1770
8. Jarrold MF (1991) Science 252:1085
9. Zhang RQ, Chu TS, Cheung HF, Wang N, Lee ST (2001) Mater Sci Eng C 16:31
10. Wang LS, Nicholas JB, Dupuis M, Wu H, Colson SD (1997) Phys Rev Lett 78:4450
11. Nayak SK, Rao BK, Khanna SN, Jena P (1998) J Chem Phys 109:1245
12. Chelikowsky JR (1998) Phys Rev B 57:3333
13. Chu TS, Zhang RQ, Cheung HF (2001) J Phys Chem B 105:1705
14. Zhang RQ, Chu TS, Cheung HF, Wang N, Lee ST (2001) Phys Rev B 64:113304
15. Hoffmann R (1988) Rev Mod Phys 60:601
16. Kim K, Suh MS, Kim TS, Youn CJ, Suh EK, Shin YJ, Lee KB, Lee HJ, An MH, Lee HJ, Ryu H (1996) Appl Phys Lett 69:3908
17. Zhang RQ, Zhao MW, Lee ST (2004) Phys Rev Lett 293:095503
18. Yu DK, Zhang RQ, Lee ST (2002) Phys Rev B 65:245417
19. Zhang RQ, Lifshitz Y, Lee ST (2003) Adv Mater 15:635
20. Mamiya M, Kikuchi M, Takei H (2002) J Cryst Growth 237:1909
21. Buffat Ph, Borel JP (1976) Phys Rev A 13:2287

Chapter 3
Stability of Silicon Nanostructures

Abstract Unsaturated silicon nanostructures are naturally unstable and can easily develop into structures of a variety of possible morphologies with coordination largely deviated from four of bulk materials. As such, numerous possible pristine silicon nanostructures including nanospheres and nanowires have been proposed in the literature, including the thinnest silicon nanowire proposed by us. Tubular silicon nanostructures are difficult to form, as revealed by comparing their electronic structure characteristics with those in bulk-like configuration and also their carbon counterparts. There can be some local minima for the tubular structures and one of a gear-like configuration is achievable at an extremely low temperature. Surface saturation of silicon nanostructures is extremely important to achieve the structural stability and delocalized electronic structures at the band edges. The surface saturation can be achieved by hydrogenation using HF-etching, which is possible due to the polarization of the Si–Si backbone if F-terminated at the surface. Hydrogen-terminated silicon nanoparticles are thermally very stable if the hydrogen coverage is more than 50 %. Hydrogenated silicon nanostructures can offer much improved chemical stability over wet oxidation over the hydrogenated bulk surface, due to the size-dependent oxidation, and thus can be used to fabricate highly stable nanodevices.

Keywords Silicon nanostructure · Stability · Nanowires · Hydrogenation · Oxidation

3.1 Thin Stable Short Silicon Nanowires

Theoretical studies of the atomic structures of SiNWs are fundamentally important for their overall properties and growth mechanism. Among the numerous theoretical studies, Menon and Richter investigated the stability of quasi-one-dimensional structures of Si using a generalized tight-binding molecular-dynamics scheme [1]. They proposed the quasi-one-dimensional structures of Si whose

R.-Q. Zhang, *Growth Mechanisms and Novel Properties of Silicon Nanostructures*   13
from Quantum-Mechanical Calculations, SpringerBriefs in Molecular Science,
DOI: 10.1007/978-3-642-40905-9_3, © The Author(s) 2014

surfaces closely resemble one of the most stable reconstructions of the crystalline Si surfaces with a core of buck-like fourfold coordinated atoms. Marsen and Sattler reported wires of 3–7 nm in diameter and at least 100 nm long [2]. The wires tend to be assembled in parallel in bundles. They proposed a fullerene-type Si_{24}-based atomic configuration for the nanowires. We have investigated the thinnest possible structures of the stacked silicon nanowires and their growth characteristics using full-potential linear-muffin-tin-orbital molecular-dynamics method [3]. Figure 3.1 shows some stable structures of selected Si clusters and wires. Si_{18}, Si_{20}, Si_{45}, and Si_{47} correspond to the stacked structures from the tricapped trigonal prisms. Si_{42} consists of the trigonal prisms. Si_{57} refers to the stacked trigonal prisms inserted among trigonal prisms by one tricapped trigonal prism. The binding energy per atom is listed below the corresponding structure. The structures stacked by the tricapped trigonal prisms are very stable compared with the other stacked structures. They can grow up to at least 26 Å. The structures stacked by the uncapped trigonal prisms are the thinnest wire structures. As the stacked layers increase, their binding energies and energy gaps tend toward saturation and zero, respectively. The structures are expected to have a saturation length larger than 30 Å. It is more important that their mixed structures can grow up to 50 Å at least. Their electronic density of states shows that their gaps are only a few tenths of an eV. The gaps decrease as the atom number increases.

Fig. 3.1 Stable structures of selected Si clusters and wires. Si_{18}, Si_{20}, Si_{45}, and Si_{47} correspond to the stacked structures from the tricapped trigonal prisms. Si_{42} consists of the trigonal prisms. Si_{57} refers to the stacked trigonal prisms inserted among trigonal prisms by one tricapped trigonal prism. The binding energy per atom is listed below the corresponding structure. Reprinted with permission from Ref. [3]. Copyright 2002, The American Physical Society

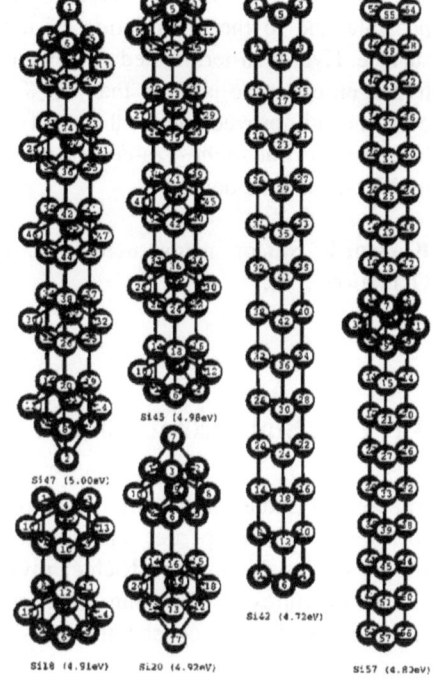

3.2 Energetic Stability of Silicon Nanotubes

One-dimensional silicon nanostructures other than SiNWs are also fundamentally important, because of the possibility to be integrated with the next generation electronics. Inspired by the formation of carbon nanotubes which is well known to be due to the ease of sp^2 hybridization of carbon, tubular silicon nanostructures have also been intensively researched in the past decade. Different from carbon, the strong sp^3 hybridization of Si does not facilitate the formation of tubular structures. Therefore, synthesis of hollow one-dimensional silicon nanotubes (SiNTs) is difficult due to the sp^3 hybridization in silicon, other than sp^2 hybridization in graphite.

In analogy to the conventional carbon nanotubes (CNTs), SiNTs had been studied theoretically by many groups [4–6]. Among them, Fagan et al. [7] established theoretical similarities between silicon and carbon nanotubes by DFT calculations and showed that the electronic and structural properties of SiNTs are similar to those of CNTs, exhibiting metallic or semiconductive behaviors depending on the structure type (zigzag, armchair, or chiral) and the tube diameter. Considering silicon's "inability" to adopt the sp^2 coordination, Seifert et al. [8] pointed out that the existence of SiNTs is doubtful and proposed that Si-based silicide and SiH nanotubes are theoretically stable and energetically viable. These structures could thus be considered as sources of silicon nanotubes, particularly in view of the existence of many layered silicides. We explored the possibility of the existence of SiNTs based on semiempirical calculations and found that SiNTs could in principle be formed with puckered surfaces under appropriate conditions [9]. The strain energy of such structures has recently been described by Barnard et al. [10]. In a more recent work, Zhang et al. [11] obtained a similar tubular structure optimized by DFT calculations. It is obvious from these studies that the calculated structures of SiNTs are sensitive to the method used.

We compared the electronic structures of a diamond nanowire, a silicon nanowire (SiNW), a CNT, and a SiNT (see Fig. 3.2) [9], which showed that carbon nanotubular structure shows efficient sp^2 hybridization and π bonding, thus allowing a high stability of the CNT structure. And not surprisingly, silicon prefers sp^3 hybridization and favors the tetrahedral diamond-like structures, thereby forming the commonly observed nanowires. Specifically, the significant difference between C and Si in their tendency to form tubular structures can be traced to the differences in the energetics and overlaps of the valence s and p orbitals of C versus Si. Due to the relatively small energy level difference, Si tends to utilize all three of its valence p orbitals, thereby resulting in sp^3 hybridization and the formation of the diamond-like nanowire structure. In contrast, the relatively large energy level difference for C implies that carbon will "activate" one valence p orbital at a time, as required by the bonding situation, giving rise, in turn, to sp, sp^2 (tube), sp^3 (wire) hybridizations. Moreover, the $\pi - \pi$ overlap in the Si = Si bond is roughly one order of magnitude smaller than the corresponding value in C = C bond in comparison with that of the C = C bond lengths. The poor $\pi - \pi$ overlaps

Fig. 3.2 Structures of the
silicon nanotube. Reprinted
with permission from
Ref. [9]. Copyright 2002,
Elsevier

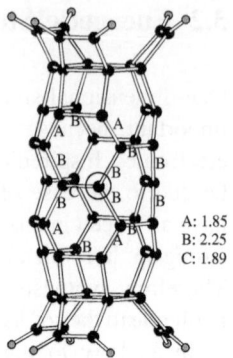

A: 1.85
B: 2.25
C: 1.89

and weak π bonding between silicon atoms give rise to a larger bond alternation (i.e., less electron delocalization) and a severely puckered structure for SiNTs.

Based on the above result, we elucidated the differences in the structures and bonding between cubic (diamond-like) and tubular nanostructures of carbon and silicon reason(s) for the hitherto unsuccessful synthesis of SiNTs. We obtained that SiNT can in principle be formed, when the dangling bonds are properly terminated. The resulting energy minimized SiNT adopts a severely puckered structure (with a corrugated surface) with Si–Si distances ranging from 1.85 to 2.25 (Å), as shown in Fig. 3.2 [9]. Such SiNT structures may serve as models for the design and synthesis of silicon nanotubes.

Our further calculations revealed that SiNTs can adopt a number of distorted tubular structures, depending on the theory and the initial models adopted [12]. In particular, "gear-like" structures shown in Fig. 3.3 with alternating sp^3-like and sp^2-like silicon local configurations are dominant for SiNTs according to our density-functional tight binding molecular dynamics simulations. The gear-like structures of SiNTs deviate notably from the smooth-walled tubes. However, they are energetically less stable than the "string-bean-like" SiNT structures previously derived from semiempirical molecular orbital calculations [9].

We further studied double-walled silicon nanotubes (DWSiNTs) with faceted wall surfaces, as shown in Fig. 3.4, and found that they are also favorable configurations [13]. They have higher energetic favorability than the conventionally adopted cylindrical configurations of single-walled silicon nanotubes (SWSiNTs). We found that the hexagonal (h-) and tetrahedral (t-) like structures of these DWSiNTs are almost energetically equivalent.

The faceted DWSiNTs though energetically more favorable than the SiNW by about 0.078 eV/atom still have high formation energies and the tendency to become nanowires at high temperature. They could be fabricated in experiments as metastable forms of Q1D silicon nanostructures.

Note that we have seen a few experimental reports on the synthesis of SiNTs, supporting the predictions described above. One of them is the preparation of crystalline SiNTs (designated as cSiNTs), which may be described as a hollow crystalline SiNW [14, 15]. While this morphology represents a new silicon

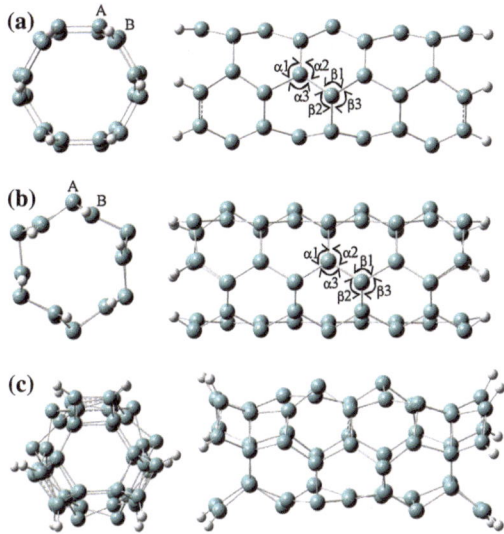

Fig. 3.3 Optimized structures of the armchair (3,3) SiNT with ab initio calculations at the HF/6-31G(d) level starting with the structures of **a** a smooth tube analogous to CNT, **b** the gear-like configuration obtained by performing MD, and **c** string-bean-like puckering obtained by using PM3. Reprinted with permission from Ref. [12]. Copyright 2005, American Chemical Society

nanotubular structure, it is very different from the conventional rolled-up graphite-like sheet structures optimized by the CNTs. More recently, Tang and coworkers [16, 17] synthesized SiNTs from silicon monoxide powder under supercritically hydrothermal conditions with moderate temperature and press. The obtained SiNTs were hollow inner pores and the interplanar spacing of silicon wall layers was 0.31 nm. Moreover, Crescenzi and coworkers [18] reported the experimental imaging of SiNTs with very thin walls and showed that the armchair SiNTs were semiconducting. Note that layered silicon systems exist in some silicides, for instance, in alkaline-earth metal silicides [19], where the silicon layers, formed by cyclohexane-like rings, are separated by metal ions. It is conceivable that the layered structure of silicon in these systems may roll up to form tubular structures analogous to the "gear-like" configurations.

3.3 Energetic Stability of Hydrogen-Terminated Silicon Nanostructures

3.3.1 Effects of Structural Saturation of Silicon Nanostructures

Porous silicon [20], as a star silicon nanostructure, has received wide attention in the scientific community due to its emission of visible luminescence. It was widely

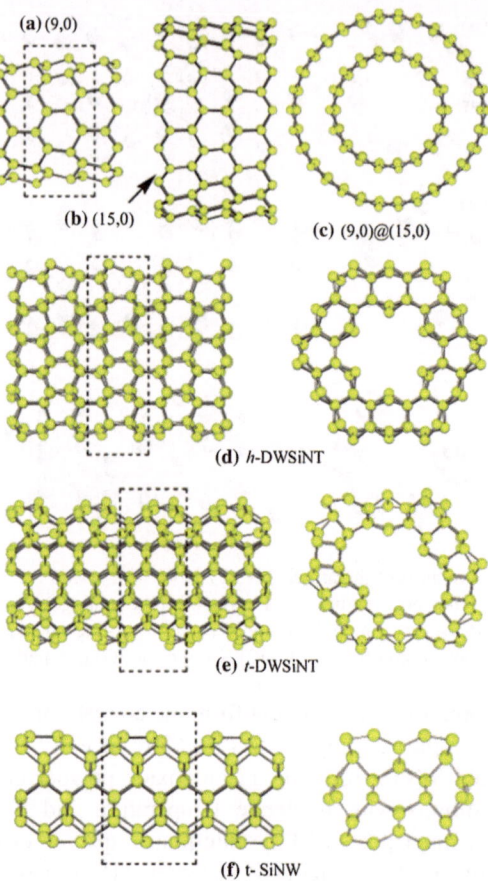

Fig. 3.4 Equilibrium configurations of SiNTs and a SiNW. Side view of **a** (9,0) and **b** (15,0) g-SWSiNTs; **c** *top view* of a cylindrical DWSiNT built by assembling (9,0) and (15,0) g-SWSiNTs with a coaxial structure; *side view* (*left*) and *top view* (*right*) of **d** h-DWSiNT, **e** t-DWSiNT, and **f** t-SiNW. Reprinted with permission from Ref. [13]. Copyright 2007, American Chemical Society

believed that the light emission arises from nanometric and crystalline silicon domains which show a quantum confinement effect [21–24]. However, other studies support different mechanisms of luminescence: several experimental and theoretical papers claimed [25–27] that both the quantum confinement and the surface (hydrogen saturation) effects are responsible; Brandt et al. [28] suggested that it may be attributable to a Si-backbone polymer, such as $Si_6O_3H_6$ (some later studies [29, 30] also supported a similar idea); Xu et al. [31] assigned luminescence to molecules attached to the Si surface; a few authors attributed it to hydrogen-related surface species such as SiH_2 and polysilanes [32, 33]; the formation of amorphous silicon has also been considered as a possible luminescent mechanism [34]. Some later papers [35–37] are very helpful in understanding the origin of the visible luminescence.

Other nanoscale silicon-based materials, such as ultrafine silicon particles [38, 39], crystallized amorphous Si:H/SiNx:H multiquantum-well structures [40], and the laser-annealed hydrogenated silicon powder produced in plasma-enhanced chemical-vapor deposition processes [41, 42] were found to also give similar visible luminescence, although their photoluminescence (PL) dynamics and energetic distribution are very different. The similarity strongly suggests that the visible luminescence be associated with some common feature of the nanosize silicon-based materials. It can be deduced that those structures can be quite different but must possess certain local ordering domains.

To prove our view and to understand the luminescent mechanism, we performed first-principles calculations of the atomic and electronic structures of a series of clusters: Si_5, Si_5X_{12}, Si_5Y_4 (X = H, F, Cl, OH; and Y = N), $Si_{17}H_n$ (n = 0, 12, 24, and 36), $Si_{29}H_{36}$, $Si_{35}H_{36}$, $Si_{41}H_{60}$, and $Si_{17}O_{12}H_{12}$. The Si_5 is a spherical cluster which has a tetrahedral T_d symmetry (two layers of silicon atoms in which a central silicon atom, Si1, is surrounded by four other surface ones, Si2). However, it does not correspond to the global minimum on the energy surface [43]. In the T_d Si_5 cluster, the dangling bonds on the surface are almost isolated and form a weak p bond with high symmetry around the surface. This highly symmetric, weak p bond enables this cluster to be maintained in a metastable structure with a Td symmetry. We found that there are several dangling bond states inside the gap from the analysis of the density of states (DOS) [44], with all the gap states being localized and coming from the surface atoms (Si2). However, saturation of the dangling bonds of the T_d Si_5 cluster with 12 hydrogen atoms led to a stable cluster, spherical Si_5H_{12}, with Td symmetry. This cluster may be considered as a crystallite and all the dangling-bond states had moved to the valence and conduction bands. The energy gap obtained is larger than the corresponding one of the Td Si_5 as shown in Table 3.1, since the saturation with hydrogen on the surface also removes the influence of dangling bonds on the bulk silicon atoms. For a larger unsaturated silicon cluster, the dangling bond may affect several layers of atoms and lead to deviations of the bonds from the bulk ones, resulting in some tail states which correspond to the deviated bonds [45] and narrowing the energy gap.

We moved on to a larger cluster, a Td Si_{17}. It may be formed by the Td Si_5 with 12 additional silicon atoms in the third layer. Two types of dangling bonds present: 12 dangling bonds isolated and the other 24 forming 12 weak-bent bonds [45]. Without saturating the dangling bonds, the structure does not form a tetrahedral Td symmetrical crystallite. Similar to the case for the Si5 cluster, the 36 dangling bonds for the Si17 cluster may be saturated with hydrogen atoms to produce a stable structure with Td symmetry. If only the 24 dangling bonds are saturated, the structure may form weak-bent bonds and leave the other 12 isolated. However, if only the 12 isolated dangling bonds are saturated, we can easily obtain a Td symmetrical crystallite. The above indicates that a key factor in obtaining a stable crystallite is the saturation of the isolated dangling bonds.

Certainly these two types of, or other similar, unsaturated dangling bonds can easily be found in a larger unsaturated cluster, and also in the nanostructured

Table 3.1 Summary of some calculated energy gaps (eV)

Cluster	Symmetry	DF[a]	PM3	Hartree–Fock		MP2 6–31G*	CIS 6–31G*
				6–31G*	6–311++G**		
Si_5^b	T_d		7.14	12.64			
Si_5	$D3\,h$		4.82	7.74			
Si_5H_{12}	$C2_v$	5.5(5.8)	6.83	12.25	11.02	12.35	10.83
Si_5H_{12}	T_d	6.3	7.28	13.44	11.57	13.47	
$Si_{17}H_{36}$	T_d	5.2	6.09	11.32			
$Si_{29}H_{36}$	T_d	4.9	5.59	10.58			
$Si_{35}H_{36}$	T_d	4.8	5.41				
$Si_{41}H_{60}$	T_d		5.29				

[a] Data of density-functional approach with self-energy corrections were estimated from Ref. [23, and references therein], and the datum in the parentheses is experimental; also cited in Ref. [23, and references therein]

[b] A metastable structure

Reprinted with permission from Ref. [46]. Copyright 1996, American Physical Society

materials. Our findings suggest that in all cases the role of the isolated dangling bonds in the stability of the structure is the same. We conclude that the cluster with dangling bonds, especially isolated ones, will easily deviate from the crystallite, or in other words, it is difficult to maintain the tetrahedral coordinated structure in the part which is affected by the dangling bonds.

Table 3.1 lists the calculated energy gaps of these clusters and others. It is seen that the energy gaps are much larger than the corresponding value of 1.12 eV of the bulk crystalline silicon, and also relatively larger than those by other theoretical calculations [21, 23, and references therein]. In general, (i) the gap becomes smaller when the size of the silicon cluster grows; (ii) the energy gap depends on the basis set and methods used; and (iii) the geometrical optimization of the cluster structure actually enlarges the energy gap. An unoptimized geometry is responsible for the tail states of the bands which may narrow the calculated gap.

Considering that F, Cl, and N or species (OH) may be present in the air, solutions, or other containments, we tested them separately as the saturator of the boundary dangling bonds for the Td Si_5 cluster. We used one F, Cl atom, or one OH group to saturate one dangling bond, whereas one N atom saturated three dangling bonds. We saw similar effects on those obtained with hydrogen as a saturator. We found that only the detailed distribution of the band states and the values of the gap shown in Table 3.2 are different. The differences are responsible for different distributions of states for a crystallite, and thus the luminescence.

We found that the oxygen atom cannot be used to saturate the boundary of the Td Si_5 cluster, since it would prefer a site near two dangling bonds However, oxygen atoms could be used for a Td Si_{17} cluster where 12 oxygen atoms were used to saturate the 24 dangling bonds which can form 12 bent bonds, and 12 hydrogen atoms saturated the 12 isolated dangling bands. We obtained similar effects and the energy gap around 5.18 eV by PM3. We concluded that the oxygen atom is also an efficient saturator for luminescence, which may play an important

Table 3.2 The calculated energy gaps (eV) for Si_5X_{12} and Si_5Y_4 (X = H, F, Cl, OH, and Y = N)

Cluster	Symmetry	PM3	HF/6-31G*
Si_5H_{12}	T_d	7.28	13.44
Si_5F_{12}	T_d	7.09	15.19
Si_5Cl_{12}	T_d	6.14	12.98
$Si_5O_{12}H_{12}$	T_d	6.51	13.13
Si_5N_4	T_d	6.13	9.11

role in high-temperature processes, e.g., annealing and fabrication. Thus, we attributed the PL induced by high-temperature annealing from a-SiOx:H film [47] to the structure crystallization with the help of oxygen. Finally, we could understand most of the reported luminescent phenomena in various silicon-based nanometric materials with the help of the study.

As we have demonstrated, there is a need to saturate the dangling bonds, especially the isolated dangling bonds of Si clusters, to maintain the nanometric and crystalline structural domains. We showed that it is indispensable for producing visible luminescence. Our calculations of the energy gap and DOS revealed that both the saturation atoms and the cluster size are responsible for the different luminescence distributions. Thus our research has linked most of the reported mechanisms of the luminescence in porous silicon, such as the confinement effect and surface species effect. Our result would be useful to understand the luminescence in any silicon-based nanostructured material.

3.3.2 Hydrogenation of Silicon Quantum Dot Surfaces

Hydrogenation is seen to be a simple method to stabilize silicon surfaces against oxidation [48]. It can be achieved simply after etching of the silicon surfaces in an aqueous HF acid followed by rinsing in H_2O [49]. Hydrogen-terminated silicon surfaces are constantly the attracting systems for intensive study due to their great technological importance in microelectronics. They are also crucial in determining the growth mode of the heteroepitaxy [50]. Nowadays, the hydrogenation of silicon surface is a controllable process with respect to the surface smoothness and the type of hydrides [51]. Ideal hydrogen termination of the Si (111) surface by monohydride existing on the entire surface could be achieved with low defect and impurity density as well as good stability [52]. However, oxidation processes still take place on a hydrogen-terminated [53] atomically flat silicon surface [52], whether produced by HF treatment or by surface passivation [54].

The chemical cleaning techniques, which can be used to achieve the hydrogenation of silicon surface, play an important role in the manufacture of integrated circuits in the semiconductor industry. In the so-called RCA cleaning technique

[55], the oxides of silicon surfaces are removed by dipping the silicon slices into an HF solution. The etched silicon surfaces were found to be passivated by H instead of F for both silicon wafer [55] and silicon nanowires [56, and references therein]. This is puzzling since a Si–F bond is almost twice as strong as a Si–H bond (bond energies: 5.7 eV vs 3.1 eV [57]). Because of the high electronegativity of F, one may expect that silicon surfaces are F-terminated in the final step of the oxide removal. Attempts have been made to elucidate the phenomenon in various theoretical and experimental studies. In 1990, for example, Trucks et al. [58] reported a mechanism of hydrogen passivation of silicon surfaces from the viewpoint of transition-state theory, by considering the hydrogen fluoride molecule as the main reactive species attacking the Si surfaces. Based on the assumption that the Si–F bonds polarize the silicon back bonds (the underlying layer) due to their highly ionic nature, this polarization was considered to facilitate the insertion of HF into the Si–Si bond, leading to fluorination of the surface silicon and hydrogenation of the underlying (second) layer silicon. Two representative transition states for F-terminated and H-terminated reactions were compared. Judging from the calculated energy barriers (1.4 eV vs 1.0 eV for the formation of F- vs H-terminated reactions), it was found that the reaction between the silicon surfaces and the incoming HF molecule is more likely to lead to the formation of H-terminated surfaces. However, in 1991, Sacher and Yelon [59] suggested a modification to the mechanism proposed by Trucks et al., arguing that, owing to steric reasons, an S_N2 attack by F- on the F-bonded Si is more likely than an attack by the HF molecule. In reply, Trucks et al. [60] pointed out that F- ions in solution are quite benign and much less reactive than HF.

While the HF etching finds new applications in silicon-based nanoscience and nanotechnology in which the etched silicon nanowire surface is evidenced to be more stable than that of a silicon wafer [56, and references therein], our theoretical study provided a view of the mechanism, and, hopefully, a better understanding, of hydrogen passivation on silicon (111) surfaces. Our calculations included the electronic structures and the comparison of bond energies and bond strengths (Mulliken bonding populations) of Si–Si, Si–F, and Si–H bonds of hydrogen- and fluorine-terminated silicon surfaces. To support the conclusions drawn from the energetic results, we also investigated the kinetics of the reactions between HF and the silicon surfaces.

Our calculations show that the bonds between surface and underlayer silicon atoms are weakened by the F-bonded species on the surface. For the trifluorine-terminated silicon surfaces, due to the polarization induced by the F-bonded moieties, the neighboring Si–Si bonds exhibit a high degree of ionic character, thereby facilitating attacks by the highly polar HF molecules. For the monofluorine-terminated silicon surface, the Si–Si bonds in the immediate vicinity of the F-bonded Si site exhibit covalent character and the calculated small bond energy suggests that these bonds will be broken more easily than the corresponding bonds on the monohydrogen-terminated silicon surfaces. Furthermore, transition-states calculations (with consideration of the solvent effect) using the cluster models Si_{10}–SiF_3, Si_9–SiF, Si_{10}–SiH_3, and Si_9–SiH to simulate the possible reactions of the Si surface with HF also suggest that H-terminated surfaces are less susceptible to HF attacks than F-terminated surfaces, in agreement with electronic structure calculations. The agreement between the thermodynamic and the kinetic results, based solely on theoretical calculations, is rather gratifying. It reinforces the well-known facts that (1) despite its behavior as a weak acid, HF is quite reactive; (2) in spite of the extraordinarily strong Si–F bond, the highly polarized Si–F can be kinetically very reactive under ionic conditions; and (3) though silicon generally forms covalent bonds (with ionic character), it usually undergoes ionic reactions and/or participates in bimolecular reactions in which bond forming and bond breaking occur simultaneously. Our computational results are in accordance with these characteristics of the silicon chemistry.

3.4 Thermal Stability of Hydrogen-Terminated Silicon Nanostructures

Theoretical studies on hydrogen-terminated Si nanoparticles have been heavily dedicated to the electronic and optical properties. For example, using a tight binding method, Ren et al. studied the density of states (DOS) and the bandgap of the hydrogenated Si nanoparticles with increasing size [61, 62]. They found that a particle with a diameter of about 4.9 nm exhibits similar DOS and bandgap to those of the bulk phase. Delley and Steigmeier studied the Si nanocrystals using the density functional approach [23, 63]. The bandgap was found to scale linearly with inversion of the particle diameters, while the luminescence intensity exhibits a strong decrease with increasing size. Delerue, Allan, and Lannoo also studied theoretically the luminescence of silicon nanoparticles and nanowires [64]. The bandgaps of the nanoparticles were found to scale as $d^{-1.39}$. The bandgap energies of the nanosized structures were in good agreement with the photon energies observed in the luminescence measurement, showing the validity of the quantum confinement model in the PL of silicon nanostructures. Wang and Zunger investigated the dependences of energy gaps and radiative recombination rates on the size, shape, and orientation of the Si nanocrystals using an empirical pseudopotential method [65]. The bandgap was found to be insensitive to the surface orientation and to the overall shape of the particles. As we mentioned in the last section, the surface saturation effects of silicon small particles are important on their electronic structures and thus determine their PL behavior [66]. All the above-mentioned calculations assumed an ideal structure of the bulk phase and terminated the surface dangling bonds with hydrogen atoms.

Among the theoretical works on structural properties of hydrogenated silicon systems, detailed structures and energetics of very small Si_nH_x (n < 16) clusters have been carried out using quantum chemical calculations [67–69]. Klein et al. systematically studied the structural and dynamic properties of a-Si:H using a tight-binding Hamiltonian constructed based on calculations with the density functional approach [70, and references therein]. It was found that Si atoms are mainly fourfold coordinated, with around 7 % fivefold coordinated atoms, while H atoms exhibit a tendency for clustering. In addition, Kratzer et al. studied the reaction dynamics of atomic hydrogen with the hydrogenated Si(001) surfaces [71], and the H_2 adsorption and desorption on Si(001) surfaces [72, 73]. Lee et al. studied the role of hydrogen using first-principles theory for Si adatom adsorption and diffusion on hydrogenated Si(001) surfaces [74]. To our knowledge, there have been very few theoretical studies relating to the detailed structures of the hydrogen-terminated silicon nanocrystals.

We have studied the structures of hydrogenated Si nanocrystals and nanoclusters using the empirical tight-binding method [75]. The hydrogenated nanocrystals are optimized using the tight-binding approach. It is shown that the structural properties of the hydrogen saturated Si nanocrystals have little size effect, contrary to their electronic properties. The surface relaxation is small in the fully hydrogen saturated Si nanocrystals. Only the atoms in the outermost two or three layers exhibit small lattice contractions of 0.01–0.02 Å. The surface relaxation is mainly dependent on the local environment, while the overall size and shape of the nanocrystals have little effect. Inside the hydrogenated Si nanocrystals, there is a very small lattice expansion of the order of 10^{-4}–10^{-3}, consistent with the X-ray diffraction measurement. For the small SimHx (m < 151) nanoclusters, simulated annealings have been performed to get the energetically favorable structures. It is found that the fully hydrogenated Si nanocrystals are the most stable structures compared to those partially hydrogenated. Removing up to 50 % of the total terminating H atoms only causes lattice distortions to the crystal structure, while the tetrahedral structures are retained. By removing more than 70–80 % of the total terminating H atoms, the clusters will evolve to more compact structures.

3.5 Chemical Stability of Hydrogen-Terminated Silicon Nanostructures

3.5.1 Surface Hydride Configuration Dependence

Hydrogen terminations are the natural product in light-emitting porous silicon [20], which could be prepared simply by electrochemical etching and has stimulated intensive studies because of its potential applications in silicon-based optoelectronic devices [76–78]. The passivation provided by hydrogen from the electrolyte is very efficient in fresh sample but unstable [79]. In the same family of nanostructured materials, one-dimensional silicon nanowire is considered more promising for its inherent quantum confinement effect in the other two dimensions [80]. Large-scale synthesis, which is always an essential requirement for wide applications of silicon nanowires has recently been achieved routinely [81, and the

references therein]. However, problems of degradation [82] and low photoluminescence (PL) efficiency [83] from the silicon nanostructures remain unsolved.
One possible solution is to passivate the silicon surface with elements other than
hydrogen. Among the existing passivation methods, nitridation is generally
believed to be of advantage [84]. However, the removal of the surface oxide layer
in order to achieve a satisfactory nitriding effect is technologically complicated.
We reported the carbidation treatment of SiNWs and achieved significantly
enhanced and stable PL [85]. An optimal carbidation and a reduced damage during
the surface treatment still require an intensive study. The problem of stability
requires additional efforts to make the promising silicon nanostructures applicable
for the widely desired nanotechnology, in addition to the advancement of conventional microelectronics. An ideally terminated, nonreactive silicon surface
could be still achieved by hydrogen, which has been the driving force for much of
the effort expended in surface science [86]. For reaching the goal, a deeper
understanding of the hydrogenated silicon surfaces with respect to their bonding
and chemical reaction with ambient gases would be crucially important.

Numerous experimental and theoretical research studies on hydrogen adsorp-
tions on clean and reconstructed Si surfaces and the related hydrogen binding
energies have been reported [87]. A systematic study on the relative bond strength
among the SiH_x (x = 1 − 3) configurations needs further pursuing, although the
study is important for figuring out a stable hydrogenated surface. Although the
silicon surface oxidation, in which water reaction may take the key role, has been
intensively studied both experimentally and theoretically [88, 89], most of the
research work was directed to the water adsorption and dissociation on clean
silicon surfaces [90].

Considering the above, a systematic investigation of the water reaction on
hydrogenated silicon cluster surfaces was conducted by us recently [91]. We first
compared the bond energies of the various hydride configurations which possibly
exist on various silicon cluster or nanowire surfaces. In conjunction with a further
study on the reaction of water molecule on them as well as on the related Si–Si
backbone, we explored the stability of hydride configurations and the origin of
surface degradation.

The hydrogen atoms on silicon cluster surfaces could be evaporated at high
temperature. The stability of monohydride configurations could be simply determined by the Si–H bond energy as there is no direct interaction between neighboring hydrogen atoms. However, at dihydride or trihydride configurations,
evaporations of hydrogen molecules may be involved. On the other hand, the
stability of hydrogenated silicon cluster surfaces may be determined by the
reactions of the surface configurations with water molecule which could exist in
the ambient environment of the silicon products.

Our calculations showed that the Si–H bond energy is simply determined by its
local configuration and is about 75.2 ∼ 76.3 kcal/mol for silicon monohydride,
77.9 ∼ 78.6 kcal/mol for silicon dihydride and 80.9 ∼ 81.6 kcal/mol for silicon
trihydride. The evaporation energies of a hydrogen molecule from the dihydride
and trihydride configurations were found slightly higher than the calculated bond

energies. However, when water molecule reacts with them, the reaction energy barriers were found to be generally smaller than 50.0 kcal/mol, much less than these bond energies. The calculated reaction barriers and heats did not show clear relationships with the bond energies. Rather, the results showed that the reaction at SiH$_2$ is the most unfavorable one while the most easy reaction may take place at the Si–Si dimer on a (2×1) reconstructed Si(001)-like configuration. Our results indicated that the degradation of hydrogenated silicon nanostructures or bulk silicon surfaces might be significantly determined by the possibility of reaction with water molecules, a hydrogenated surface covered by dihydride configurations being the most inert case.

For all the reactions considered here, the hydrogen abstraction and addition channels, the water addition reaction on the reconstructed configuration was the easiest, showing that the Si–Si backbone can be easily broken due to its considerably strained structure. Moreover, the reaction of water addition onto the hydrogenated Si(111)-like configuration which also leads to the Si–Si bond breaking may possibly occur. It can be expected that the hydrogen abstraction reaction and the addition reaction of water on the unreconstructed silicon configurations would be two competitive reaction channels. Nevertheless, a dihydride configuration is shown to be the most stable one among the considered cases.

3.5.2 Size-Dependent Oxidation

Since there are often size effects appearing for properties such as energy gaps when the size of silicon structures reaches a nanometer, we explored further the size effect of oxidation and, thus, the chemical stability. The study of reaction of water molecules on various hydrogenated silicon clusters can relate the stability with the local hydride configuration and shed light on a way to achieve stable, unreactive hydrogenated silicon structures. Eventually, our result has revealed the size dependence on reactivity of water molecule interacting with small silicon clusters [92].

The results for the water reaction on the dihydride (solid lines) and trihydride (dashed lines) configurations are depicted in Fig. 3.5. A similar reactivity trend has been revealed with the reaction energetics in a wide range of temperatures, in particular at a temperature below 400 °C in all cases. For the reaction on the dihydride configuration, the reaction rate calculated for the medium cluster shows a considerably large increase with respect to that for the smallest cluster, but could only be slightly increased when the cluster size is further increased.

The reactivity trend of the various hydrogenated silicon clusters with varied cluster size in reaction with a water molecule can also be seen from the analysis of their frontier orbitals. Early studies [93], and references therein] have established that the overlap between the highest-occupied molecular orbital (HOMO) of one molecule and the lowest-unoccupied molecular orbital (LUMO) of another would determine the nature of chemical reaction. A smaller energy difference between

Fig. 3.5 Total rate constants (cm^3 mol^{-1} s^{-1}) at 1 atm, 300 K for the water reaction on the dihydride (*solid lines*) and trihydride (*dashed lines*) configurations. Reproduced with permission from data published in Ref. [92]. Copyright 2002, American Institute of Physics

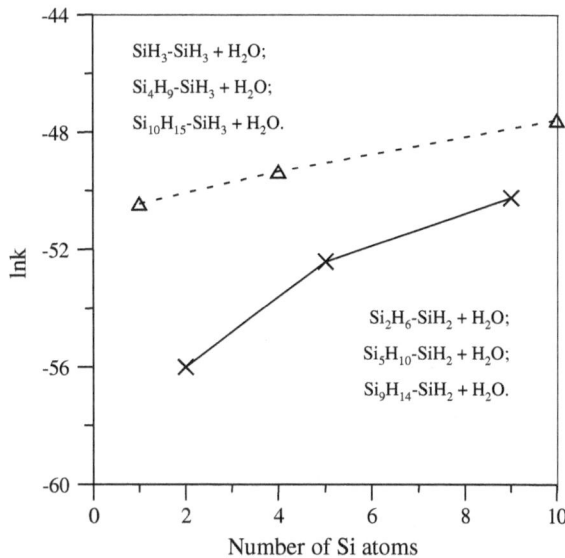

the HOMO of one molecule (electron donor) and the LUMO of the other (electron acceptor) would indicate a more favorable reaction taking place. For the reactions at the same kind of hydrogenated silicon configurations, the energy differences (LUMO$_W$–HOMO) obtained from the HOMOs and LUMOs (see Table 3.3) of the considered systems (the water molecule and the hydrogenated clusters) illustrate a similar trend to that revealed by the above study on reaction energetics (also summarized in Table 3.3) and rate constants. Such a trend is closely related to the HOMOs and LUMOs of the individual hydrogenated silicon clusters, for which, when increasing the cluster size, their HOMOs generally move up and LUMOs down, resulting in decreased energy gaps, as shown in Table 3.3. As the size effect of the energy gap is well known for silicon clusters when the size changes in the range of nanometer, the correlation of the reactivity with the well-documented size effect of the energy gap would provide important implications to the size-dependent reactivity.

Similar to the size effect of the energy gap, the size effect of reactivity would be expected to take effect once the local structure reaches a nanometer. Since the energy gap of silicon clusters has been found to stabilize at a certain value when the cluster size is sufficiently large [23, and references therein], it is expected that the reactivity here and reaction rate constant illustrated in the previous part for a fairly large silicon cluster would also stabilize at a certain value for a given temperature and pressure. Because of the restricted number of cluster models with varied sizes considered in our work due to the limits of computational resources, such a trend could not be clearly shown in our results, but is reasonably true.

Our finding of size-dependent reactivity would provide an important scientific basis for realizing nonreactive, stable hydrogenated silicon structures. Such

Table 3.3 LUMO, HOMO, and energy gaps for the various substrate clusters determined at HF/6–31G** level of calculation and their reactivities with a water molecule (HOMO$_W$ = −0.4976 Hartree; LUMO$_W$ = 0.2153 Hartree) shown by LUMO$_W$–HOMO energy difference

Substrate clusters	HOMO (Hartree)	LUMO (Hartree)	LUMO–HOMO (eV)	Reaction with water		
				LUMO$_W$–HOMO (eV)	Energy barrier (kcal/mol)	Reaction heat (kcal/mol)
Si$_2$H$_6$–SiH$_2$	−0.3830	0.1190	13.7	16. 3	47.3	−13.5
Si$_5$H$_{10}$–SiH$_2$	−0.3678	0.1051	12.9	15.9	45.5	−14.8
Si$_9$H$_{14}$–SiH$_2$	−0.3653	0.0930	12.5	15.8	44.5	−14.9
SiH$_3$–SiH$_3$	−0.4022	0.1400	14.8	16.8	44.6	−15.3
Si$_4$H$_9$–SiH$_3$	−0.3799	0.1143	13.4	16.2	43.4	−15.3
Si$_{10}$H$_{15}$–SiH$_3$	−0.3519	0.0869	11.9	15.4	42.6	−16.0

The reaction barriers and reaction heats determined at MP2/6–31 + G**//HF/6–31G** level of calculation for the corresponding reactions are also included. Reprinted with permission from Ref. [92]. Copyright 2002, American Institute of Physics

nonreactive, stable structures could be achieved by the formation of uniform nanosized structures (should be less than 10 nm, at which point the size effect becomes significant). The oxidation found for porous silicon is probably due to the poor uniformity and insufficiently small size of its nanostructures. The oxidation resulting from some reactive local structures may lead to the propagation of the oxidation, due to the oxidation at the backbone. Once the silicon structure is well hydrogenated and the surface reconstruction is removed, the reaction taking place on the backbone of the silicon structure would become less possible. Thus, to obtain stable, nonreactive hydrogenated silicon structures, it is important to achieve unreconstructed structures, all uniformly covered by silicon dihydride, in particular, with the structural dimensions in nanometers. Our finding indicates the possibility of fabricating highly stable nanodevices based on hydrogenated silicon structures.

References

1. Menon M, Richter E (1999) Phys Rev Lett 83:792
2. Marsen B, Sattler K (1999) Phys Rev B 60:11593
3. Li BX, Zhang RQ, Cao PL, Lee ST (2002) Phys Rev B 65:125305
4. Kang JW, Hwang HJ (2003) Nanotechnology 14:402
5. Kang JW, Byun KR, Hwang HJ (2004) Modelling Simul Mater Sci Eng 12:1
6. Bai J, Zeng XC, Tanaka H, Zeng JY (2004) PNAS 101:2664
7. Fagan SB, Barierle RJ, Mota R, da Silva AJR, Fazzio A (2000) Phys Rev B 61:9994
8. Seifert G, Kohler Th, Urbassek HM, Hernandez E, Frauenheim Th (2001) Phys Rev B 63:193409
9. Zhang RQ, Lee ST, Law CK, Li WK, Teo BK (2002) Chem Phys Lett 364:251
10. Barnard AS, Russo SP (2003) J Phys Chem B 107:7577
11. Zhang M, Kan YH, Zang QJ, Su ZM, Wang RS (2003) Chem Phys Lett 379:81
12. Zhang RQ, Lee HL, Li WK, Teo BK (2005) J Phys Chem B 109:8605

13. Zhao MW, Zhang RQ, Xia Y, Song C, Lee ST (2007) J Phys Chem C 111:1234
14. Sha J, Niu JJ, Ma XY (2002) Adv Mater 14:1219
15. Teo BK, Li CP, Sun XH, Wong NB, Lee ST (2003) Inorg Chem 42:6723
16. Tang YH, Pei LZ, Chen YW, Guo C (2005) Phys Rev Lett 95:116102
17. Chen YW, Tang YH, Pei LZ, Guo C (2005) Adv Mater 17:564
18. Crescenzi MD, Castrucci P, Scarsilli M, Diociaiuti M, Chaudhari PS, Balasubramanian C, Bhave TM, Bhoraskar SV (2005) Appl Phys Lett 86:231901
19. Janzon KH, Schafer H, Weiss A, Anorg Z (1970) Allg Chem 372:87
20. Canham LT (1990) Appl Phys Lett 57:1046
21. Wang X, Huang D, Ye L, Yang M, Hao P, Fu H, Hou X, Xie X (1993) Phys Rev Lett 71:1265
22. Schuppler S, Friedman SL, Marcus MA, Adler DL, Xie Y-H, Ross FM, Harris TD, Brown WL, Chabal YJ, Brus LE, Citrin PH (1994) Phys Rev Lett 72:2648
23. Delley B, Steigmeier EF (1993) Phys Rev B 47:1397
24. Read AJ, Needs RJ, Nash KJ, Canham LT, Calcott PDJ, Qteish A (1992) Phys Rev Lett 69:1232
25. Li K-H, Tsai C, Shih S, Hsu T, Kwong DL, Campbell JC (1992) J Appl Phys 72:3816
26. Robinson MB, Dillon AC, George SM (1993) Appl Phys Lett 62:1493
27. Lee S-G, Cheong B-H, Lee K-H, Chang KJ (1995) Phys Rev B 51:1762
28. Brandt MS, Fuchs HD, Stutzmann M, Weber J, Cardona M (1992) Solid State Commun 81:307
29. Lavine JM, Sawan SP, Shieh YT, Bellezza AJ (1993) Appl Phys Lett 62:1099
30. Xiao Y, Heben MJ, McCullough JM, Tsuo YS, Pankove JI, Deb SK (1993) Appl Phys Lett 62:1152
31. Xu ZY, Gal M, Gross M (1992) Appl Phys Lett 60:1375
32. Tsai C, Li K-H, Sarathy J, Shih S, Campbell JC, Hance BK, White JM (1992) Appl Phys Lett 59:2814
33. Tsai C, Li K-H, Kinosky DS, Qian R-Z, Hsu TC, Irby JT, Banerjee SK, Tasch AF, Campbell JC, Hance BK, White JM (1992) Appl Phys Lett 60:1700
34. Vasquez RP, Fathauer RW, George T, Ksendzov A, Lin TL (1992) Appl Phys Lett 60:1004
35. Prokes SM, Carlos WE, Glembocki OJ (1994) Phys Rev B 50: 17093
36. Schuppler S, Friedman SL, Marcus MA, Adler DL, Xie Y-H, Ross FM, Chabal YJ, Harris TD, Brus LE, Brown WL, Chaban EE, Szajowski PF, Christman SB, Citrin PH (1995) Phys Rev B 52:4910
37. Brus LE, Szajowski PF, Wilson WL, Harris TD, Schuppler S, Citrin PH (1995) J Am Chem Soc 117:2915
38. Furukawa S, Miyasato T (1989) Superlattices Microstruct 5:317
39. Morisaki H, Ping FW, Ono H, Yazawa K (1991) J Appl Phys 70:1869
40. Chen KJ, Huang XF, Xu J, Feng D (1992) Appl Phys Lett 61:2069
41. Costa J, Roura P, Sardin G, Morante JR, Bertran E (1994) Appl Phys Lett 64:463
42. Roura P, Costa J, Sardin G, Morante JR, Bertran E (1994) Phys Rev B 50:18124
43. Raghavachari K (1986) J Chem Phys 84:5672
44. Zhang RQ, Wang JJ, Dai GC, Wu JA, Zhang JP, Xing YR (1989) Chin J Semicond 10:327
45. Zhang RQ (1989) Solid State Commun 69:681
46. Zhang RQ, Costa J, Bertran E (1996) Role of structural saturation and geometry in the luminescence of silicon-based nanostructured materials. Phys Rev B 53:7847
47. Lin C-H, Lee S-C, Chen Y-F (1993) Appl Phys Lett 63:902
48. Chabal YJ (1984) Phys Rev B 29:3677
49. Chabal YJ, Higashi GS, Raghavachari K, Burrows VA (1989) J Vac Sci Technol A 7:2104
50. Sumitomo K, Kobayashi T, Shoji F, Oura K, Katayama I (1991) Phys Rev Lett 66:1193
51. Higashi GS, Becker RS, Chabal YJ, Becker AJ (1991) Appl Phys Lett 58:1656
52. Higashi GS, Chabal YJ, Trucks GW, Raghavachari K (1990) Appl Phys Lett 56:656
53. Hirashita N, Kinoshita M, Aikawa I, Ajioka T (1990) Appl Phys Lett 56:451
54. Ikeda H, Hotta K, Furuta S, Zaima S, Yasuda Y (1069) Jpn J Appl Phys 1996:35
55. Kern W, Puotinen DA (1970) RCA Rev 31:187

56. Zhang RQ, Lifshizh Y, Lee ST (2003) Adv Mater (Weinheim, Ger) 15:635
57. Lide DR (2003) Handbook of chemistry and physics, 83rd edn. CRC, Boca Raton, pp 9–55 (CRC, Boca Raton, FL, 2002–2003)
58. Trucks GW, Krishnan R, Higashi GS, Chabal YJ (1990) Phys Rev Lett 65:504
59. Sacher E, Yelon A (1991) Phys Rev Lett 66:1647
60. Trucks GW, Krishnan R, Higashi GS, Chabal YJ (1991) Phys Rev Lett 66:1648
61. Ren SY, Dow JD (1992) Phys Rev B 45:6492
62. Hirao M, Uda T (1994) Surf Sci 306:87
63. Delley B, Steigmeier EF (1995) Appl Phys Lett 67:2370
64. Delerue C, Allan G, Lannoo M (1993) Phys Rev B 48:11024
65. Wang LW, Zunger A (1994) J Phys Chem 98:2158
66. Zhang RQ, Costa J, Bertran E (1996) Phys Rev B 53:7847
67. Onida G, Andreoni W (1995) Chem Phys Lett 243:183
68. Miyazaki T, Uda T, Štich I, Terakura K (1996) Chem Phys Lett 261:346
69. Meleshko V, Morokov Yu, Schweigert V (1999) Chem Phys Lett 300:118
70. Klein P, Urbassek HM, Frauenheim Th (1999) Phys Rev B 60:5478
71. Kratzer P (1997) J Chem Phys 106:6752
72. Kratzer P, Hammer B, Nørskov JK (1995) Phys Rev B 51:13432
73. Kratzer P, Pehlke E, Scheffler M, Raschke MB, Höfer U (1998) Phys Rev Lett 81:5596
74. Lee SM, Lee YH, Kim NG (2000) Surf Sci 470:89
75. Yu DK, Zhang RQ, Lee ST (2002) Structural properties of hydrogenated silicon nanocrystals and nano-clusters. J Appl Phys 92:7453
76. Cullis AG, Canham LT (1991) Nature (London) 353:335
77. Sham TK, Jiang DT, Couithard I, Lorimer JW, Feng XH, Tan KH, Frigo SP, Rosenberg RA, Houghton DC, Bryskiewicz B (1993) Nature (London) 363:331
78. Buda F, Kohanoff J, Parrinello M (1992) Phys Rev Lett 69:1272
79. Tischler MA, Collins RT, Stathis JH, Tsang JC (1992) Appl Phys Lett 60:639
80. Morales AM, Lieber CM (1998) Science 279:208
81. Zhang RQ, Lifshitz Y, Lee ST (2003) Adv Mater 15:639
82. Cooke DW, Bennett BL, Farnum EH, Hults WL, Sickafus KE, Smith JF, Smith JL, Taylor TN, Tiwari P (1996) Appl Phys Lett 68:1663
83. Yu DP, Bai ZG, Wang JJ, Zou YH, Qian W, Fu JS, Zhang HZ, Ding Y, Xiong GC, You LP, Xu J, Feng SQ (1999) Phys Rev B 59:R2498
84. Daami A, Bremond G, Stalmans L, Poortmans J (1998) J Lumin 80:169
85. Zhou XT, Zhang RQ, Peng HY, Shang NG, Wang N, Bello I, Lee CS, Lee ST (2000) Chem Phys Lett 332:215
86. Becker RS, Higashi GS, Chabal YJ, Becker AJ (1917) Phys Rev Lett 1990:65
87. Nachtigall P, Jordan KD, Janda KC (1991) J Chem Phys 95:8652
88. Weldon MK, Queeney KT, Gurevich AB, Stefanov BB, Chabal YJ, Raghavachari K (2000) J Chem Phys 113:2440
89. Teraishi K, Takaba H, Yamada A, Endou A, Gunji I, Chatterjee A, Kubo M, Miyamoto A, Nakamura K, Kitajima M (1998) J Chem Phys 109:1495
90. Ranke W, Xing YR (1997) Surf Sci 381:1
91. Zhang RQ, Lu WC, Zhao YL, Lee ST (2004) J Phys Chem B 108:1967–1973
92. Zhang RQ, Lu WC, Lee ST (2002) Appl Phys Lett 80:4223
93. Hoffmann R (1988) Rev Mod Phys 60:601

Chapter 4
Novel Electronic Properties of Silicon Nanostructures

Abstract Due to their finite size and the significant localization of their electrons upon excitation, silicon nanostructures in excited states undergo severe relaxation and thus show significant Stokes shift at a diameter less than 1.5 nm. The effect is much reduced at a larger size due to the improved structural rigidity and also due to the delocalization of the excited electrons. The latter can also be achieved by elongating the nanostructure in a certain direction. One-dimensional silicon nanowires present energy band structures of strong orientation and size dependences. In particular, <112> silicon nanowires always have indirect bandgaps if the cross-sectional aspect ratio of the (110) and (111) facets is smaller than 0.5. At a larger aspect ratio, the bandgap becomes direct. The bandgap can also be tuned by applying external stress, to direct one with a compression up to 5 %, but keeps indirect under a tensile stress. For two-dimensional silicon sheets, the possibility to tune the bandgap between indirect and direct is very high, by carefully controlling the magnitude and direction of strain application, very effective for engineering the electronic band structure of silicon nanostructures.

Keywords Silicon quantum dots · Silicon nanowires · Excited state · Stokes shift · Emission · Surface · Silicon sheets · Band dispersion · Stress · Straining

4.1 Excited State Relaxation of Hydrogen-Terminated Silicon Nanodots

The optical properties of silicon quantum dots have long been a subject of intensive experimental and theoretical research due to their potential applications in advanced electronics and optoelectronic devices. Yet, a systematic comparison between theoretical and experimental results for these systems has been hindered by the lack of structure-specific experimental data. The enormous amount of experimental data reported on silicon dots [1–9] still require an understanding of

R.-Q. Zhang, *Growth Mechanisms and Novel Properties of Silicon Nanostructures from Quantum-Mechanical Calculations*, SpringerBriefs in Molecular Science, DOI: 10.1007/978-3-642-40905-9_4, © The Author(s) 2014 the actual structures of quantum dots or particles. At present, spectroscopic data linked to specific particle structures are only available for a few small Si_nH_m clusters. In the study of the properties of macroscopic samples, all experimental data for larger particles and quantum dots represent the "ensemble average" values for a large number of particles that do not contain precise information about the geometries, surface conditions, or the exact number of atoms in clusters. The structures of hydrogenated silicon dots have a very strong influence on their optical properties, implying that a comprehensive comparison of experimental and theoretical results requires the precise knowledge of particle structures. Such information is available only for very small Si_nH_m clusters. Besides, the origin of strong room temperature PL in porous silicon needs to be correctly assigned to the appropriate structure for a proper understanding and practical utilization of the mechanism. Theoretical studies are very useful in validating the experimental structures and may be employed in tandem with experiments for a more fruitful investigation. Some important computational findings on hydrogenated Si quantum dots by us are discussed in the following paragraphs.

We use Fig. 4.1 to pictorially depict the fluorescence phenomenon [10, 11]. Prior to its photonic excitation, the H-SiQDs remains in its ground state at (1) corresponding to its energy minimum. Upon photoabsorption, it raises itself to its excited state at (2), with absorption energy equal to the difference between energies at (2) and (1). The (2) usually does not correspond to its minimum in the

Fig. 4.1 Schematic energy diagram of the ground state and the first excited singlet state of silicon nanoparticles. Reprinted with permission from Ref. [10]. Copyright 2007, American Chemical Society

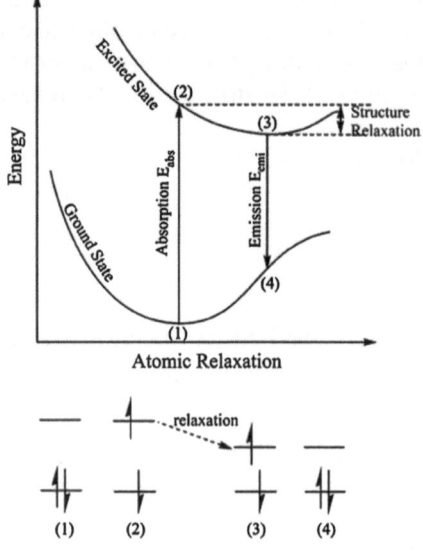

excited state. It undergoes a nonradiative structural organization or relaxation to reach its minimum at (3) at the excited state. Then, the dot undergoes a radiative transition from (3) to (4) and emits a radiation of energy equal to the difference between the energies at (3) and (4). The difference between absorption and emission energies is termed as Stokes shift.

The absorption energy is a property inherent to the ground, equilibrium state of the system and is also known as the energy gap or difference between the edges of the conduction and valence bands. The energy gap is a synonym for the difference in energy between the highest occupied molecular orbital (HOMO) and the lowest unoccupied molecular orbital (LUMO) of the cluster. Different theoretical methodologies have been used to estimate the absorption energy of hydrogenated Si nanoparticles. Delerue, Allan, and Lannoo studied theoretically the luminescence of silicon nanoparticles and nanowires [12]. The energy gaps of the nanoparticles were found to scale as $d^{-1.39}$. The absorption energies of the nanosized structures were in good agreement with the photon energies observed in the luminescence measurement, showing the validity of the quantum confinement model in the PL of silicon nanostructures. Wang and Zunger investigated the dependences of energy gaps and radiative recombination rates on the size, shape, and orientation of the Si nanocrystals using an empirical pseudopotential method [13]. The energy gap was found to be insensitive to the surface orientation and to the overall shape of the particles. All the above calculations assumed an ideal structure of the bulk phase and terminated the surface dangling bonds with hydrogen atoms.

Our optimized hydrogen-terminated spherical Si nanoparticles with diameters of 3.0, 2.5, 2.4, 2.2, 2.1, 2.0, 1.8, 1.7, 1.4, and 1.2 nm, respectively, by an empirical tight-binding approach [14] show energy gaps of a linear behavior with the inverse diameter (D^{-1}), which is consistent with previous results of tight-binding and density-functional calculations [15–18], in conformity with the quantum confinement effect. The gap-diameter relationship is fitted as $E_g = 0.45 + 3.33/D$, where E_g is in eV and D in nm. For Si particles of 2.0 and 3.0 nm, the gaps are 2.12 and 1.56 eV, respectively, in agreement with previous tight-binding calculations [17]. In our studies, the energy gap saturates to the bulk value at a large cluster size consisting of about 200 Si atoms. The change in the absorption energy (E_{abs}, in eV) versus the diameter of silicon nanoparticles (d_0, in nanometers) could be fitted to the equation: $E_{abs} = 7.156 \exp(-d_0/1.032) + 1.773$ (eV) [19], using the self-consistent-charge density-functional tight-binding (SCC-DFTB) methodology [20, 21].

Emission energy is a chief characteristic of the excited, nonequilibrium state of the dots and is therefore, far more complex compared to the absorption energy. It is controlled by the relaxation dynamics of the dot at its excited state. It is desirable to minimize Stokes shift for the purpose of maximizing its quantum yield or efficiency in PL applications. This underscores the need for clarity in the understanding of the organizational dynamics of the dots at their excited states. Photoexcitation causes destabilization of the dot as the system gains energy with respect to its stable ground state, a concomitant to the creation of an exciton (a virtual or a quasi-particle) or an electron–hole pair. Alternatively, an electron is promoted from HOMO, a bonding state of lower energy to LUMO, which is either a nonbonding state of higher energy or an antibonding (orbital) state of higher energy. The destabilization of the dot upon photoexcitation may show up as a shape deformation or through the weakening or stretching of some bonds in the dot. As discussed earlier in a previous section, any deformation in shape from the tetrahedral structure narrows the energy gap. Symmetry in structure, which is T_d here in this case, leads to degeneracy in electronic states. When the tetrahedral symmetry is broken upon shape deformation at excited state, the degeneracy at LUMO energy of ground state is lifted and a state decouples out to appear in the HOMO–LUMO gap of the ground state, which in turn narrows the gap at excited state. An equivalent argument to account for this reduction in energy gap is as follows. As the bond length between two atoms increases, the bond is weakened. As a result, the electronic cloud in the bonding region shifts by some proportion to the two atoms connected by the bond, thereby inducing some degree of charge localization on these two atoms. Corresponding to this localization, a state appears in the energy gap, thereby reducing the gap at excited state. The greater the magnitude of structural distortion or charge localization at excited state, the higher will be the downshift of LUMO at excited state with respect to the LUMO at ground state and correspondingly, lower emission energy and higher Stokes shift. Our ab initio calculations have shown that the electronic structure of the hydrogen atoms in the H-SiQDs remains unaffected upon photoabsorption or excitation. On reaching the excited state, it is rather the electronic cloud around the silicon atoms of the cluster which undergoes reorganization. This is not surprising because the energetic distribution of the electronic cloud of the dot arises solely out of the electronic structure of the Si atoms held together in tetrahedral T_d symmetry or order. When it comes to electronic structure of the dot, the role of hydrogen is passive, as it merely saturates the dangling bonds of Si and thereby facilitates the maintenance of the T_d symmetry of the Si core.

4.1.1 Size Dependency of Emission Properties

In contrast to a great deal of theoretical studies [22–32] on the energy gap or absorption energy, studies on the emission gap are rather scarce, largely due to the lack of an efficient excited state geometric optimization method. However, a clear understanding of the excited state properties and its dynamics is of utmost necessity as the light emission properties of silicon nanoparticles is intimately related to it.

Similar to the studies on absorption energy, it is useful to discuss the variation of emission energy with particle size. Williamson and coworkers [24, 33] studied the light emission of ideal hydrogenated clusters by using the density-functional theory (DFT) with local density approximation and quantum Monte Carlo (QMC) and found that, similar to absorption energy, the Stokes shift is extremely sensitive to particle size. Other groups [34, 35] also reported a similar tendency in the Stokes shift. Since the size dependence of the energy gap and the Stokes shift has been widely accepted for silicon nanoparticles, a similar trend for the emission property was naturally accepted. However, it was reported [35] that, being different from the absorption energies, the emission energies of the hydrogen-terminated, saturated nanoclusters with less than 35 silicon atoms showed a slow and nonmonotonic change between 2.6 and 0.5 eV rather than a rapid increase with size decrease. Consistent with experiments, the energy gaps of silicon nanostructures were found blue-shifted from the infrared to the visible region as the size was reduced. Still, numerous controversies exist between theoretical and experimental results due to the use of different methodologies.

The diversities in the results are mainly due to (i) the uncertain particle size in the measurement, (ii) the surface impurity of the hydrogenated silicon particles, and (iii) the approximations of the contemporary approaches for studying the excited state properties. We used the self-consistent-charge density-functional tight-binding (SCC-DFTB) method for structural optimization to examine the ground state properties of silicon nanoparticles. A time-dependent linear response extension of the DFTB scheme (TD-DFTB) [36] was applied to study the excited states. The basis sets of numerically described s, p, and d atomic orbitals for silicon and an s orbital for hydrogen were used in all calculations. To obtain the excitation energies, a coupling matrix, which gives the response of the potential with respect to a change in the electron density, was built. The use of some approximations, such as a γ approximation, facilitates a high computational efficiency of the TD-DFTB approach. Thus, it is possible to study a nanoscale system containing several hundred atoms using the TD-DFTB method. After optimizing the excited state geometry by calculating the energy gradient, the emission spectrum was calculated.

The accuracy of the TD-DFTB was evaluated by comparing the results of TD-DFTB with that of time-dependent DFT (TD-DFT) at the B3LYP/6–311G* level (Refs. [40, 41]). Table 4.1 shows that the energy gaps from TD-DFTB are generally slightly lower than those from TD-DFT. For the small particle Si_5H_{12}, the energy gap is overestimated in the TD-DFT (B3LYP) calculation, while a value predicted by the TD-DFTB is lower than but closer to the experimental value (6.5 eV) [37]. The absorption energy of $Si_{17}H_{36}$ at the TD-DFTB level is even smaller than that of TD-DFT, possibly due to the incompact structure that involves too many –SiH_3 groups. Furthermore, the gross absorption features of the TD-DFTB and TD-DFT results for the other two larger particles, $Si_{29}H_{36}$ and $Si_{35}H_{36}$ are similar and in excellent accordance with the results of the frequently referred second-order perturbation theory (MP2) [38, 39]. These data are even better than the ones reported using QMC [24] and gradient-corrected Perdew–Burke–Ernzerhof functional, GGA-PBE [35]. For the light emission properties, these available theoretical methods present even larger discrepancies than in the absorption property predictions of small particles such as Si_5H_{12} for which the TD-DFTB result is much smaller than that of TD-DFT, due to the use of minimal basis. The discrepancy becomes much reduced as the size of the particle increases, as is

Table 4.1 Calculated optical gaps (in eV) for several small hydrogenated silicon nanoparticles

	$Si_5H_{12}^a$	$Si_{17}H_{36}$	$Si_{29}H_{36}$	$Si_{35}H_{36}$
TD-DFTB[b]	6.40 (2.29)	4.47 (2.40)	4.42 (2.57)	4.37 (2.89)
TD-DFT[b]	6.68 (2.96)	5.04 (2.54)	4.66	4.48
TD-DFT[c]	6.66	5.03	4.53	4.42
MP2[d, c]			4.45	4.33
QMC[e]			5.3	5.0
DFT/GGA[f]			3.65	3.56

[a] The experimental absorption value is 6.5 eV [37]. Data in parentheses are the corresponding emission energies [19]
[b] Our work [19]
[c] Ref. [38]
[d] Ref. [39]
[e] Ref. [24]
[f] Ref. [14]. Reprinted with permission from Ref. [19]. Copyright 2007, American Institute of Physics

evident in the calculation of $Si_{17}H_{36}$ for which the emission energies at the TD-DFTB and TD-DFT levels are 2.40 and 2.54 eV, respectively.

Although the excited state optimizations of $Si_{29}H_{36}$ and $Si_{35}H_{36}$ using TD-DFT method are beyond our computational capability, their corresponding emission energies from TD-DFTB (2.57 and 2.89 eV) are still in close agreement with the TD-DFT/GGA-PBE calculations (2.29 and 2.64 eV), respectively [35]. The comparison indicates that the TD-DFTB method is capable of providing results, which are close to that of first-principles DFT calculations. Furthermore, the TD-DFTB results are accurate enough for studying the optical properties of silicon nanoparticles with a significantly reduced computational cost.

Table 4.2 shows our calculated absorption and emission energies. It can be seen that, as with previous reports [24, 38], the absorption energies decrease remarkably from 6.40 to 2.81 eV with increasing size, and the decrease becomes smaller when the number of silicon atoms exceeds 87 (around 1.5 nm in diameter).

Table 4.2 Calculated absorption energies (E_{abs}) and emission energies (E_{emi}) of silicon nanoparticles with different diameters (d_0)

Particle	d_0 (nm)	E_{abs} (eV)	E_{emi} (eV)
Si_5H_{12}	0.45	6.40	2.29
$Si_{17}H_{36}$	0.98	4.47	2.40
$Si_{29}H_{36}$	1.03	4.42	2.57
$Si_{35}H_{36}$	1.09	4.37	2.89
$Si_{59}H_{60}$	1.36	3.72	3.18
$Si_{75}H_{76}$	1.41	3.51	3.12
$Si_{87}H_{76}$	1.48	3.47	3.25
$Si_{123}H_{100}$	1.74	3.11	3.04
$Si_{147}H_{100}$	1.76	3.08	2.94
$Si_{199}H_{140}$	2.00	2.81	2.76

Reproduced with permission from Ref. [10]. Copyright 2007, American Chemical Society

Fig. 4.2 Absorption and emission energies and Stokes shift of silicon particles (ranging from $Si_{26}H_{32}$ to $Si_{199}H_{140}$) versus the number of silicon atoms in them. Reprinted with permission from Ref. [10]. Copyright 2007, American Chemical Society

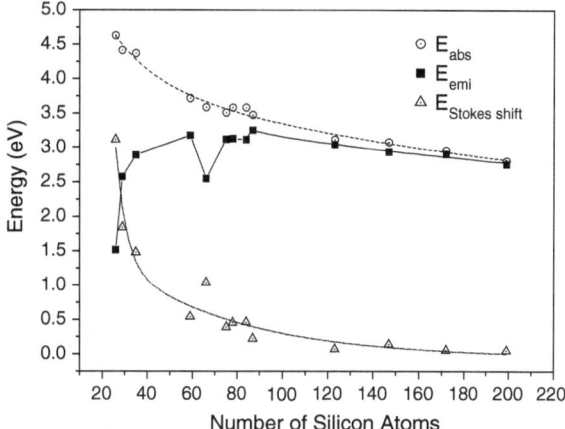

Spherical hydrogenated Si clusters of sizes ranging from 5 to 199 Si atoms were studied in our work to probe into the size dependence of optical properties of Si particles. Particles having a diameter larger than 1.5 nm (or 87 number of Si atoms) act in conformity to the quantum size effect, i.e., the emission energy decreases monotonically with the increase in the cluster size, as shown in Fig. 4.2. The emission properties of particles smaller in size than 1.5 nm were found to be at variance with the quantum confinement effect model. Below this critical particle size, the emission energies were found to decrease nonmonotonically with the decrease in particle size, as seen in Fig. 4.2. These trends have been confirmed recently by first-principles TDDFT calculations [42].

The anomalous behavior of particles below this threshold size of 1.5 nm is due to the following reasons. Small particles are weakly held together by their constituent atoms due to their low cohesive energy. Cohesive energy increases with size and as a result, particles rigidify with size. As a consequence, the response of larger particles to external stimuli (e.g., photonic excitation) is considerably different from that of smaller particles. The large particles are structurally more resistant to external excitation (e.g., photoexcitation) than the smaller ones. In other words, the ability of smaller particles to combat the destabilizing impact of the photoexcitation is less than that of the large particles.

In the ground state of silicon nanoparticles, the Si–Si bond lengths are around 2.34–2.38 Å, which is close to that in the bulk crystal.

Since the inner Si–Si bonds are slightly longer (i.e., weaker) than the surface Si–Si bonds, the former would be prone to further weakening upon photoexcitation. The weak bond(s) which can be looked upon as defects in the nanoparticle are most susceptible to external excitation. In other words, the weak/stretched bonds act as exciton traps or centers for optical activity. A silicon atom connecting to the central silicon atom shifts away from its original position due to excited state relaxation, leading to an elongated Si–Si bond, as shown in Figs. 4.3a and 4.4b. The bond stretching (i.e., weakening) occurs by a larger magnitude for a Si_5H_{12}

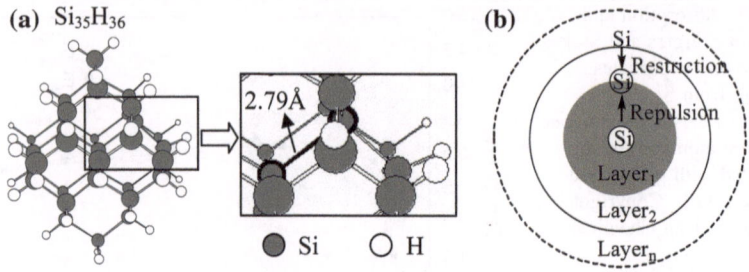

Fig. 4.3 Schematic diagrams representing structural relaxation: **a** Structure of $Si_{35}H_{36}$ in its excited state, **b** structure relaxation diagram of general spherical silicon nanoparticles. Reprinted with permission from Ref. [19]. Copyright 2007, American Institute of Physics

Fig. 4.4 **a** Comparison of Si–Si bond lengths in ground states (*black lines*) and excited states (*red lines*). The solid lines correspond to the maximal bond lengths, while the dashed lines correspond to the minimal bond lengths. **b** Schematic geometrical relaxation diagram of the $Si_{66}H_{64}$ in excited state. Reprinted with permission from Ref. [10]. Copyright 2007, American Chemical Society

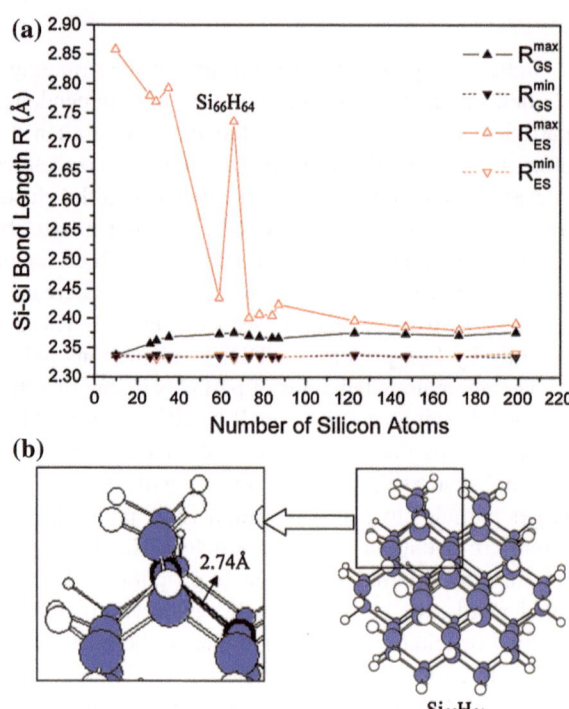

cluster upon excitation due to its smaller size and lower cohesion. This leads to a strong localization of electronic cloud on the atoms of the bond and hence, a significant lowering of LUMO level at excited state with respect to its ground state, as depicted in Fig. 4.5. For larger particles, the structural rigidity of the outer layers of atoms restrains the stretching of this bond, as portrayed in Fig. 4.3b. This bond stretching wanes as the rigidity of the particles heightens with size.

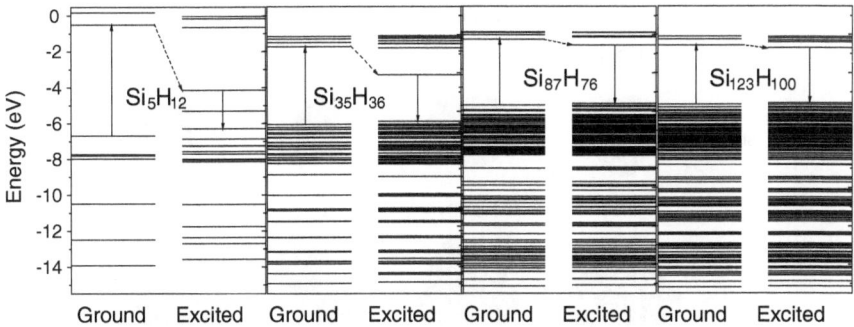

Fig. 4.5 Calculated energy levels of valence orbitals of several silicon nanoparticles in their ground and excited states. Reprinted with permission from Ref. [19]. Copyright 2007, American Institute of Physics

Consequently, the lowering of the LUMO level at excited state is reduced with size as can be seen in Fig. 4.5. This in turn shows up as decreasing Stokes shift with increasing particle size in Fig. 4.2. The key to minimizing Stokes shift is to minimize charge localization or shape deformation at excited state.

Figure 4.6 shows the frontier orbitals at both ground and excited states. The closer the resemblance of the orbitals, the closer would be the corresponding energy levels. As the shift of HOMO level in the excited state relative to its ground state is small, the orbitals of HOMO at both ground and excited states are similar in appearance. The same does not hold for LUMO for small particle sizes, as the LUMO at the excited state downshifts considerably. For larger particles, the orbital features of the LUMOs at ground and excited state get closer. The energy levels become quasi-continuous with the increase in particle size and approach the bulk behavior for larger particles, as shown in Fig. 4.5.

4.1.2   Effects of Dimerization of Si–Si Surface Bonds and Its Effects

As illustrated above, external excitation targets the weak bonds in the particle, which are considered as exciton traps or centers inducing optical activity. Correspondingly, it will be immensely useful to choose the centers of optical activity by a suitable manipulation of the particle structure or composition in order to gain direct control over its optical activities.

Since the centers for optical activities or the exciton traps are typically found to lie at the center of the cluster, the extent of shape deformation of the cluster at excited state will be considerable in response to bond stretching occurring at the center. (For sufficiently small clusters, it could even dissociate the cluster apart, as discussed in a later sub-section.) This will in turn induce a large downshift of

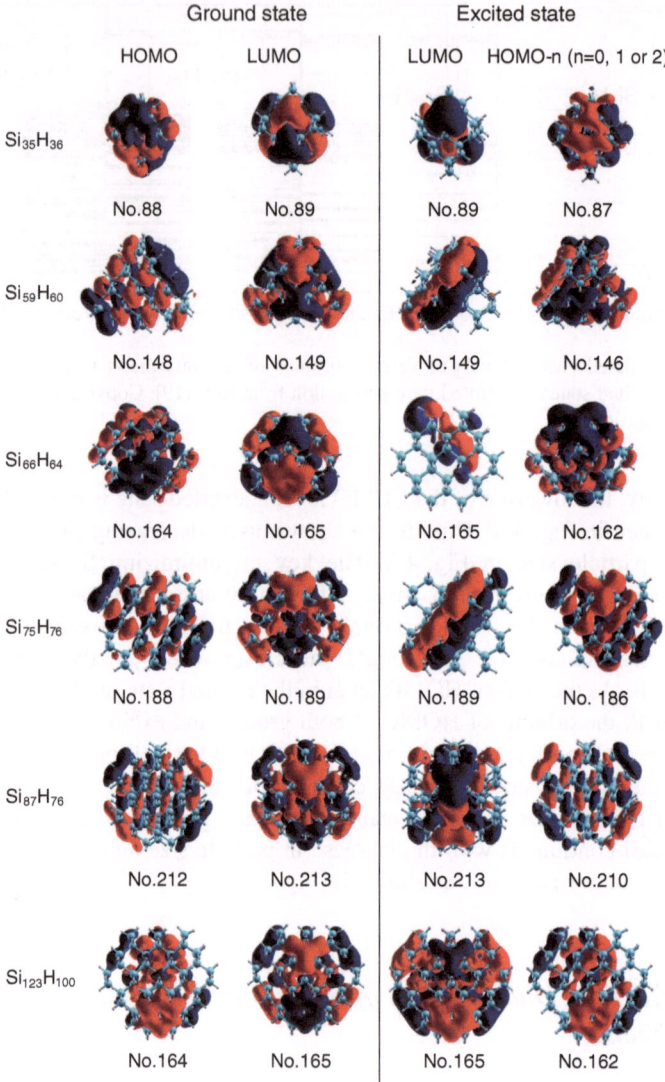

Fig. 4.6 Isosurfaces of wave functions of silicon nanoparticles in ground and excited states plotted at the same isovalue. Reprinted with permission from Ref. [10]. Copyright 2007, American Chemical Society

LUMO at excited state, reducing the energy gap and the emission energy, in accordance with the earlier discussion. Consequently, the Stokes shift will undesirably increase. However, if the exciton traps or centers for optical activity are designed to lie on the surface, the excitation will then not affect much the shape of

H-SiQDs at excited state. As a result, the Stokes shift will be controlled or min-imized and the PL efficiency of the H-SiQDs will be maximized. Partial dehy-drogenation [14] is not found to affect the stability of the tetrahedral structure of the Si nanocrystals, although the surface reconstructs a little on dehydrogenation.

Yet, dehydrogenation and the concomitant surface reconstruction obviously change the electronic structure and property of the dot. It is very crucial to understand the change in electronic structure caused by the dehydrogenation in order to be able to utilize this understanding to the application needs of the dot. Unfortunately, direct experimental evidence of the reconstructed surface, partic-ularly the predicted Si–Si dimers, is still scarce, so theoretical simulations play an important role in attempts to understand the photophysics of these systems. Using density-functional and quantum Monte Carlo methods, Puzder et al. [43] predicted that the highly curved surfaces of reconstructed silicon nanoparticles would dra-matically reduce energy gaps and decrease excitonic lifetimes, which would explain the variations in the photoluminescence spectra of colloidally synthesized nanoparticles and the observed deep gap levels in porous silicon. As already mentioned, most theoretical studies concerned with surface reconstruction still focus on absorption energy [43–47] because simulations of the emission process are computationally highly demanding.

The effect of surface reconstruction in a series of hydrogen-terminated silicon particles of sizes up to 2 nm with varying number of Si–Si dimers for models ranging from perfect bulk-like to fully reconstructed configurations were investi-gated [48] in our work, as the surface dimers were suggested to act as exciton traps [44, 49] or centers for unusual optical activity. Selective partial dehydrogenation by removal of dihydrides as sketched in Fig. 4.7 induces dimerization on the surface as it introduces a weak π bond between these two Si atoms. These weak π bonds then act as exciton traps. Positive formation energies for all sizes displayed in Fig. 4.7a imply destabilization of Si nanoparticles upon complete surface reconstruction, but the extent of this destabilization is reduced if the sizes of the particle increase. The formation energy, E_f is directly proportional to the ratio of the number of surface dimers to its volume. For $Si_{29}H_{36}$, the formation energy is 1.76 eV when it dissociates to $Si_{29}H_{24} + 6H_2$, while E_f is only 0.37 eV in the reaction $Si_{199}H_{140} \rightarrow Si_{199}H_{92} + 24H_2$ due to the larger particle size. The high formation energy of $Si_{66}H_n$ (2.19 eV) is attributed to the larger dimer ratio, as compared to the particles of other sizes, $Si_{29}H_n$ and $Si_{78}H_n$ because the former has 12 Si–Si surface dimers while the latter two has only 6. Both the E_f curves of $Si_{29}H_n$ and $Si_{199}H_n$ in Fig. 4.7b, c show that (within a narrow range of fluctuation) the formation energies generally decrease first and then increase again if more and more dihydrides (SiH_2) are replaced by Si–Si dimers. The most stable particles for Si_{29} and Si_{199} cores are the ones capped by 32 and 116 H atoms, respectively; which correspond to 2 and 12 Si–Si dimers on the particle surfaces. It turns out that bulk-like configurations may not necessarily be the most stable configurations; as for instance, the partially reconstructed clusters for Si_{29} and Si_{199} cores override the stability of their fully hydrogenated counterparts.

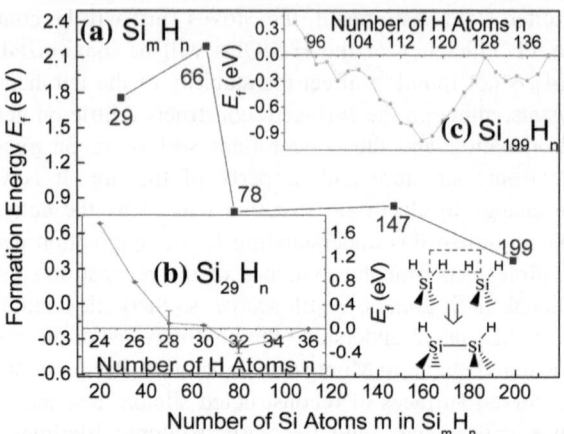

Fig. 4.7 **a** Formation energies of silicon nanoparticles Si_mH_n versus number of Si atoms, n for completely reconstructed structures. The inset illustrates dimerization on the surface; **b** Formation energies of $Si_{29}H_n$ versus the number of H atoms n ($n = 24–36$). *Error bars* correspond to different isomers; **c** Formation energies of $Si_{199}H_n$ versus the number of H atoms n ($n = 92–140$). Please note that for each particle, only one of several possible isomers was studied. Reprinted with permission from Ref. [48]. Copyright 2008, American Institute of Physics

Because of lower stability of fully reconstructed Si clusters relative to their bulk counterparts, the former exhibits lower absorption energy and energy gap. While the absorption energy of bulk-like particles red-shifts with increasing system size as discussed in the preceding section, this trend does not hold for the small reconstructed particles, especially for those smaller than $Si_{78}H_n$. This is attributed to the dot structure adapted upon surface reconstruction. As mentioned earlier, the Si–Si bond lengths in bulk-like particles in ground state are about 2.33–2.37 Å. However, in reconstructed particles, especially in small size particles, the bond lengths of surface Si–Si dimers are longer than 2.4 Å, even stretching to 2.44 Å in $Si_{66}H_{40}$ and 2.46 Å in $Si_{29}H_{24}$, which results in structures with lower formation energy and smaller energy gaps in comparison to bulk-like species. If the size further increases ($>Si_{78}H_{52}$), the bond lengths of Si–Si dimers tend to be close to the Si–Si bond lengths in bulk-like particles in ground state, so the absorption energies follow the usual size dependence of bulk-like particles. Since the dimer bonds are already stretched (i.e., weakened) in the ground state in particles ($<Si_{78}H_{52}$), these surface dimer bonds would be susceptible to further weakening on photoexcitation than the other bonds. The surface dimers, in this way, act as exciton traps or centers for optical activity.

Figure 4.8 shows that the emission energies were found to differ significantly from the absorption energies, giving rise to huge Stokes shifts of the order of 2 eV. In contrast to the absorption energies, the emission energies for reconstructed particles surprisingly increased with system size. This counterintuitive behavior continued until a size of 2 nm ($Si_{199}H_{92}$) was reached. From this point onwards, the spectrum was expected to follow the quantum confinement picture with negligible Stokes shifts, similar to the evolution of the spectra of bulk-like particles. This is because the geometrical distortion at excited state occurs on the surface of Si–Si dimer bonds, which fully agrees with an earlier prediction. The outer atomic layers and their structural rigidity which restrains the bond stretching at excited state in the bulk-like particles is completely absent in the reconstructed particles.

For $Si_{29}H_{24}$, one of the surface Si–Si dimers stretches from 2.46 to 2.98 Å, while for $Si_{78}H_{52}$, one of the surface Si–Si dimers stretches from 2.43 to 2.92 Å. On the other hand, the strongest structural distortions for both $Si_{66}H_{40}$ and $Si_{147}H_{76}$ occur at a next-neighbor surface dimer Si–Si, rather than at the dimer itself. In Fig. 4.8, the unusual optical properties of $Si_{66}H_{40}$ can be attributed to its large number of Si–Si dimers, which leads to high surface tensile stress and an accompanying small optical gap. For bulk-like particles, in contrast, the structural distortion takes place primarily in the first-neighbor silicon atoms at the core center. As the size increases beyond 1.5 nm ($>Si_{87}H_{76}$), the emission energies in the bulk-like particles show a size dependence conforming to the quantum size effect model because the strengthened structural rigidity contributed by the outer layers lessens the Stokes shifts with respect to the corresponding partially saturated reconstructed particles where outer layers, which could possibly combat surface tensile stress, are missing. Therefore, one has to optimize both the size and the extent of hydrogenation of H-SiQDs simultaneously in order to maximize their PL efficiency in technological applications.

Thus, it is clear that both the size and degree of hydrogenation needs to be controlled in order to have the desired wavelength of luminescence from the Si quantum dots.

Fig. 4.8 The calculated absorption and emission energy of bulk-like (*solid lines*) and completely restructured (*dashed lines*) Si nanoparticles. The *inset* shows the structures of $Si_{29}H_{24}$ and $Si_{66}H_{40}$ with the bond lengths of the strongest elongated Si–Si bond. The bond length after excitation is displayed in parenthesis. Reprinted with permission from Ref. [50]. Copyright 2009, American Institute of Physics

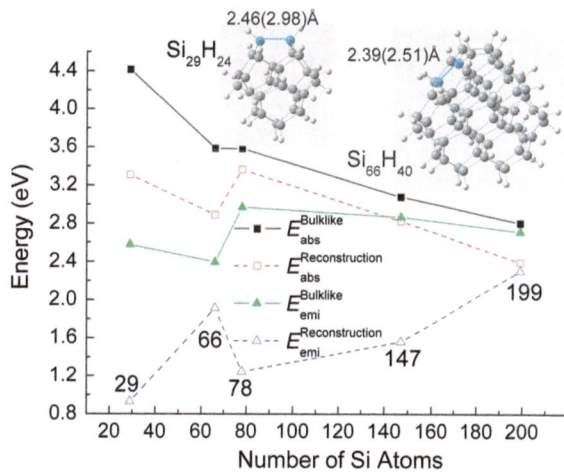

4.1.3 Exciton Traps in 1D Si Nanorods

It is interesting from both fundamental and technological perspectives to explore optoelectronic properties in other shapes or structures as well, as Si-based materials are designable in a range of shapes and sizes, based on their need or purpose. To that end, the evolution of the optical properties from small quantum dots to nanorods with large aspect ratio was investigated [51] and will be discussed in this sub-section. The short nanorods with lengths below 2 nm show localized excitons and the formation of self-trapped excitons, while excited state relaxation has little effect on longer structures which exhibit delocalized excited states, as shown in Fig. 4.9.

Figure 4.10 shows that the variation of the absorption and emission energy with the length of the nanowire displays a trend akin to that of the hydrogenated Si-QDs. Correspondingly, the Stokes shift is large for small lengths of nanorods for large-scale structural relaxation until the heightened structural rigidity at longer lengths counteracts structural relaxation or distortion at excited state.

4.2 Direct–Indirect Energy Band Transitions of Silicon Nanowires by Surface Engineering

Hydrogen-terminated SiNWs are particularly interesting and technologically important, as they are promising materials for device applications. The electronic band structure of SiNWs is known to depend strongly on wire diameter, showing a

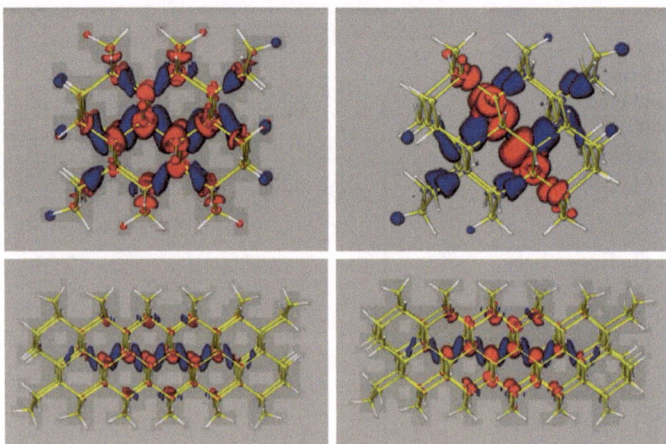

Fig. 4.9 Absolute values of the molecular orbitals, HOMO (*blue*) and LUMO (*red*) for both the optimized structures at ground state (*left*) and excited state (*right*) of $Si_{48}H_{50}$ (*up*) and $Si_{96}H_{86}$ (*down*), which are models of <110> SiNW with diameter $d = 0.84$ nm. The plot corresponds to an isovalue of 0.001. Reprinted with permission from Ref. [51]. Copyright 2009, American Chemical Society

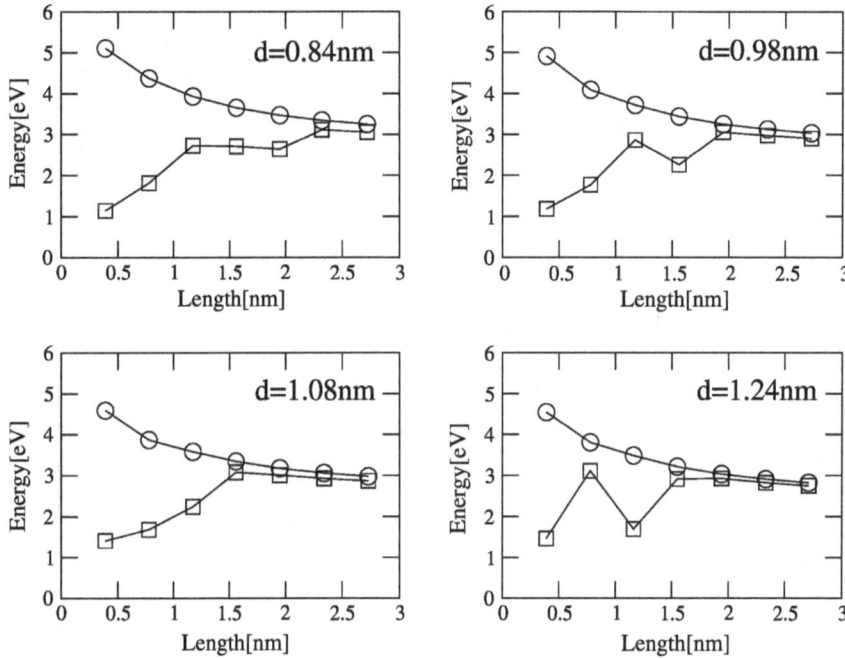

Fig. 4.10 Variation of TD-DFTB-calculated absorption energies (*circles*) and emission energies (*squares*) for <110> SiNW with different diameter and lengths. Reprinted with permission from Ref. [51]. Copyright 2009, American Chemical Society

unique trend of energy gap width increase with decreasing diameter, considerably deviating from the bulk value [52], as revealed for <100> SiNWs by Read and coauthors [53] using first-principles pseudopotential theory, and for <111> SiNWs by Zhao et al. [52] using density-functional theory (DFT) in the local-density approximation (LDA). An indirect-to-direct energy gap transition was found for < 111 > SiNWs at a diameter of less than 2.2 nm [52], with a very small difference (<0.05 eV) between the direct and indirect gaps. However, recent LDA calculation by Scheel et al. [54] on <112> SiNWs bounded with high-index (113) and low-index (111) facets showed indirect energy bandgaps for all the <112> SiNWs even at a small size (0.8 nm). Our recent results of the <112> SiNWs enclosed by low-index facets also predicted size-dependent indirect gaps [55]. The <112> SiNWs appear to have distinctly different electronic structures from SiNWs of other orientations, which warrants further investigations.

The possibility of inducing indirect-to-direct bandgap transition in silicon nanowires (SiNWs) by changing wire diameter is already well known and documented. Interestingly, it has been shown that for <112>-oriented SiNWs indirect-to-direct bandgap transition can be tuned simply by changing the cross-sectional shape of the wire or the cross-sectional aspect ratio of the (111) and (110) facets that enclose the wire [56], instead of changing the wire diameter. The cross-sectional aspect ratio must be smaller than 0.5 in order to maintain a direct bandgap, indicating the importance of the (110) facet. The finding was obtained in a theoretical study of the surface effect on the electronic band structures of <112> SiNWs using DFT with the GGA/LYP functional [56]. A unique trend of indirect-to-direct energy bandgap transition exists in the <112> SiNWs, which is predominantly controlled by the (110) facets. For the cross-sectional aspect ratio of (111) to (110) facet smaller than 0.5, <112> SiNWs have a direct bandgap, indicating that the wire cross-section shape is important in tuning the electronic band structures of SiNWs. The finding has significant implications in the application of SiNWs in optoelectronics.

We use Fig. 4.11 to describe the structure feature of a <112> SiNW. The projection view presents an n × m grid. Accordingly, the structure is denoted as AnBm. The labels "A" and "B" represent the (111) and (110) facets enclosing the SiNW, while the numbers "n" and "m" denote the numbers of grids in the projected view. For all the <112> SiNWs, the top of the valence band was always located at the Γ point, while the bottom of the conduction band was somewhere very close to X between Γ and X. We quantified the difference between the indirect bandgap and the direct bandgap by defining a Δ as the energy difference between the energy of the conduction band bottoms at Γ and the energy of the conduction band minimum at the k point near X point (see Fig. 4.12). By so doing, clearly a positive Δ represents an indirect bandgap while a negative Δ a direct bandgap.

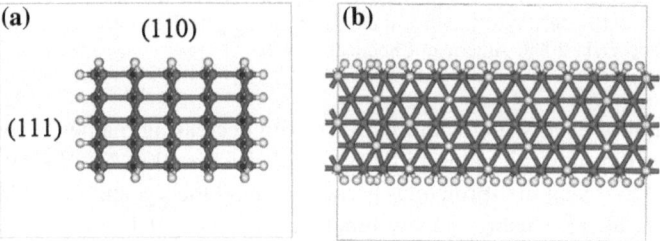

Fig. 4.11 a Top view and **b** side view of a [112]-oriented SiNW. Reprinted with permission from Ref. [55]. Copyright 2008, IOP Publishing

Fig. 4.12 Band structure of [112] SiNW. The energy difference of the conduction minimum at Γ and X point is labeled as Δ. Reprinted with permission from Ref. [55]. Copyright 2008, IOP Publishing

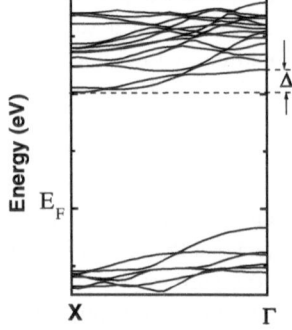

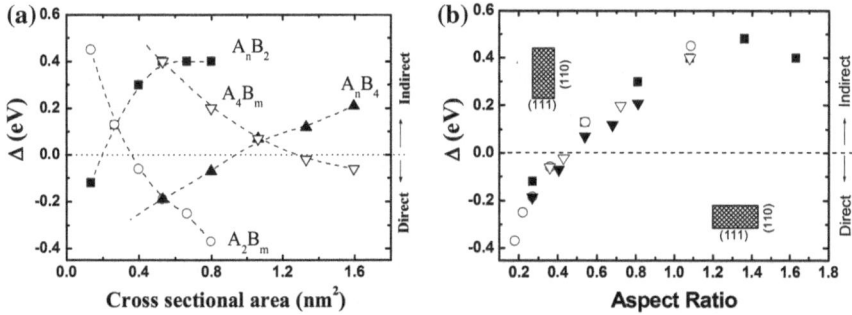

Fig. 4.13 a Energy difference (Δ) versus cross-sectional area of [112] SiNWs, where AnBm ("A" denotes the (111) facet and "B" the (110) facet) is used to distinguish the different series of SiNWs; and **b** Δ versus the area ratio of (111) to (110) side facets, with shaded rectangles representing the cross-section of <112> SiNW. The horizontal line at zero is a reference to manifest the transition between direct and indirect electronic bandgap. Reprinted with permission from Ref. [56]. Copyright 2008, American Institute of Physics

For the <112> SiNWs with $n = 1$–6 in AnB2, e.g., with the (110) facets fixed but the (111) facets variable in dimension, the band structure change versus n is shown in Fig. 4.13a with the filled squares. The Δ increases with increasing cross-sectional area; Δ is -0.12 eV at $n = 1$ and changes to 0.40 eV at $n = 6$. The line connecting the data points crosses the $\Delta = 0$ reference line at a cross-section area of 0.2 nm^2, indicating a direct-to-indirect bandgap transition. Note that the <111> SiNW undergoes a change from indirect-to-direct bandgap as wire diameter decreases, and the transition takes place at a considerably larger diameter of 2 nm [52]. Our calculations show that <112> SiNWs also undergo similar transition, but at a much smaller diameter. This indicates an orientation dependence of the energy gap transition.

The Δ values are also obtained for $n = 2$–6 in the AnB4 series, and are shown in Fig. 4.13a with filled triangles. Again, the (110) facet is fixed while the (111) facet is left variable. Here, the curve linking the filled triangles is almost a straight line. Δ also increases with wire size, but more slowly than the AnB2 series. It shows that direct-to-indirect bandgap transition also occurs, and it is a common feature of both the series of <112> SiNWs. Furthermore, although their cross-sectional areas are similar, the AnB4 shows a lower Δ value than that of the AnB2, indicating a higher possibility of obtaining a direct bandgap in AnB4 than in AnB2 SiNWs.

For the <112> SiNWs with the (111) facet being fixed but the (110) facet variable, which are denoted as A2Bm and A4Bm, the Δ values labeled by empty circles and empty inverted triangles in Fig. 4.13a all decrease with increasing wire size. The intersections of the $\Delta = 0$ line with the two curves joining the empty circles and empty inverted triangles, respectively, indicate that SiNWs undergo an indirect-to-direct bandgap transition with increasing wire size. Note that the SiNWs of other orientations studied in the previous works all showed an indirect-to-direct transition only at a sufficiently small size. In sharp contrast, our finding suggests that large-diameter <112> SiNW might become a direct gap material. This finding is remarkable and shows the unique property of <112> SiNWs in electronic band structures. Moreover, the significant difference between the A2Bm and the A4Bm series lies in the smaller Δ of the former, at the similar size. This means that the larger the (111) facet, the more difficult it is for the indirect-to-direct transition to occur. In particular, when the (111) facet is far larger than the (110) facet, the SiNW may most likely show an indirect bandgap.

The opposite trend of Δ versus cross-sectional area for AnB2 and A2Bm as shown in Fig. 4.13a is particularly interesting. As the diameter of SiNWs decreases, AnB2 undergoes an indirect-to-direct transition, whereas A2Bm a direct-to-indirect transition. For a pair of SiNWs with similar cross-sectional areas, e.g., the A1B2 and A2B1 SiNWs, the electronic band structures are totally different; the former showing a direct bandgap whereas the latter an indirect bandgap. This indicates that size is not a good index that determines bandgap transition, the cross-sectional aspect ratio between the (111) and (110) facet plays a crucial role in determining the direct or indirect nature of the bandgap.

Following Xu et al. [57], we use the cross-sectional aspect ratio to denote the cross-section shape. The variation of Δ versus cross-sectional aspect ratio is shown in Fig. 4.13b. The labels in the figure are the same as those in Fig. 4.13a; i.e., filled squares represent AnB2, filled triangles AnB4, empty circles A2Bm, and inverted triangles A4Bm. Remarkably, the datum points in Fig. 4.13b converge onto a single line, indicating that irrespective of SiNWs types Δ values follow the same trend with aspect ratio. The dependence of electronic band structure on cross-sectional aspect ratio is clearly revealed. When the ratio is 0.5, Δ is about 0, indicating the beginning of bandgap transition. This important finding implies that the electronic band structure of SiNWs could be varied by controlling the cross-sectional aspect ratio. In other words, the electronic band structure of SiNWs could be tuned simply by changing the cross-sectional shape. However, according to our previous study, the (111) facet of hydrogen-terminated SiNWs is energetically more favorable than the (110) facet [58]. Therefore, it might be difficult to achieve as-grown SiNWs with predominantly (110) side facets. It is expected that SiNWs with predominantly (110) side facets could be obtained by post treatments such as preferential etching in experiments.

To reveal the physical mechanism of such an interesting dependence, we studied surface effect on the electronic structures of SiNWs. By examining the projected density of states (PDOS), we found that the features of the PDOSs of the surface atoms and those of the core atoms were different, as shown in Fig. 4.14. Specifically, the PDOSs at the bottom of the conduction band of the atoms on the (110) facet were lower than those of the atoms on the (111) facet. Thus, the bottom of the conduction band for the total density of states (TDOS) is mainly determined by the contributions from the atoms on the (110) facet. As the aspect ratio of cross-section increases, the ratio between the areas of (111) and (110) facets decreases. As a result, the DOS at the bottom of conduction band increases, making it more

Fig. 4.14 Projected density of states (PDOS) for atoms at different positions. Solid line shows PDOS for a silicon atom in (111) side facet, dash line for central silicon atom, and dot line for atom in (110) side facet. Reprinted with permission from Ref. [56]. Copyright 2008, American Institute of Physics

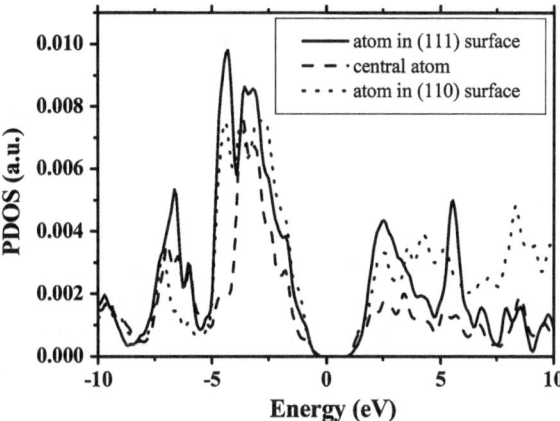

probable to become a direct bandgap. It means the (110) facets play a determinant role in the bandgap transition of SiNWs, thus making the cross-sectional aspect ratio the dominant factor for gap transition.

4.3 Tuning Energy Bands of 1D and 2D Silicon Nanostructures by Straining

4.3.1 Direct–Indirect Energy Band Transitions of 1D Silicon Nanowires

The exploration of the possibility of inducing indirect-to-direct transition in silicon is interesting and significant to develop applications of SiNWs, as indirect band structure and consequential weak light emission hamper silicon applications in optoelectronics. An interesting but less explored possibility is to tune the bandgap by introducing structural deformation.

Structural deformation of materials is well known to cause electronic structure and dipole moment changes. To offset dipole moment change, the material develops surface charges and thus voltage, leading to the well-known phenomenon named piezoelectric effect. In general, a piezoelectric effect could be extended to the changes in electrical property and thus in electronic property, in response to external compression or expansion. This effect could be observed in a variety of solid materials, and has been widely used in applications. It is expected that the electronic property change could be achieved in nanosized materials. Recently, strain has been used to enhance mobility in planar Si metal oxide semiconductor field-effect transistors [59]. There have been a few attempts to tailor the optical properties of SiNWs by strain engineering [60, 61]. Strain arises naturally during the synthesis of NWs: compressive radial strain and tensile axial strain in the SiNWs is found to occur due to the facile formation of an amorphous SiO_2 layer (1–2 nm thick) on its surface upon its exposure to air [62]. It would be interesting to investigate the effect in SiNWs at atomic level and if the electronic energy band could be tuned from indirect-to-direct under mechanical deformation.

Our calculations showed that [112] SiNWs are distinctly different from [111] and [110] SiNWs at small sizes. Specifically, the latter wires become direct bandgap materials, while the former remains indirect at a diameter less than 2 nm. Consequently, [112] SiNWs seem less useful for light absorption and emission applications. Therefore, it would be desirable to tune the energy band of [112] SiNWs so as to expand their applications, particularly in optoelectronic devices. To this end, we have further explored the possibility of tuning electronic band structures by applying external mechanical stress. Consequently, we found that axial compression or expansion could efficiently alter the electronic energy bands near the Fermi energy level and eventually result in an indirect-to-direct transition [63].

We adopted a wire model along [112] direction that possesses the same local structure as that of bulk silicon. The SiNWs have a rectangular cross-section and are surface saturated with hydrogen atoms. By manually modifying the extents of axial compression/expansion, the electronic energy band of SiNWs under different strain fields were calculated. Figure 4.15 shows the energy bands of [112] SiNWs with different lattice constants: (a) compressed 8 %, (b) compressed 5 %, (c) compressed 2.5 %, (d) no change, and (e) expanded 2.5 %. The densities of states (DOS) of cases (a) and (e) are provided in Fig. 4.15, to ensure that the band is not correlated with any localized states. Strikingly, 8 % lattice compression (a) shows typical direct bandgap, whereas 2.5 % lattice expansion (e) shows indirect bandgap, indicating that strain field modifies the energy band and induces a direct-to-indirect transition in SiNWs.

Figure 4.15 shows that the top of the valence band always remains at Γ point, whereas the bottom of the conduction band changes with axial tension. Specifically, one of the lowest conduction bands obviously moves downward under

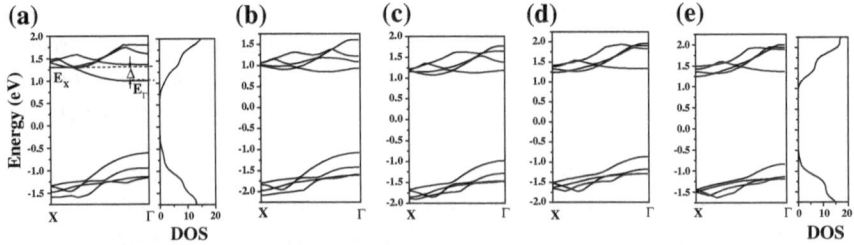

Fig. 4.15 Electronic energy bands of SiNWs with different percentage of lattice constant compression or expansion. **a** 8 % compression, **b** 5 % compression, **c** 2.5 % compression, **d** no change, and **e** 2.5 % expansion. The densities of states (DOSs; the peak width for broadening is 0.2 eV) of (**a**) and (**e**) are also provided. Reprinted with permission from Ref. [63]. Copyright 2007, American Institute of Physics

Fig. 4.16 The energy difference, Δ, in conduction band bottom at Γ and X points (E_Γ–E_X in Fig. 4.15) versus axial change in lattice constant. Reprinted with permission from Ref. [63]. Copyright 2007, American Institute of Physics

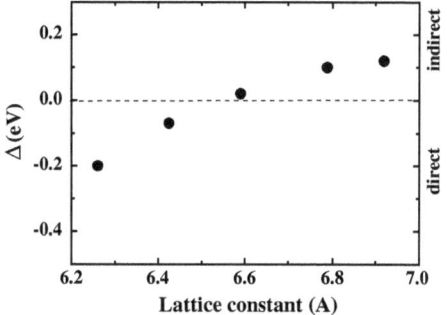

compression but lifts upward under expansion. Given that this band is not correlated with any localized states as indicated in the DOS (see Fig. 4.15a, e, it is clear that the bandgap is direct in the case of large compression but indirect in the case of expansion. The nature of bandgap can be conveniently described by a Δ value, defined as the energy difference in the bottom edge of conduction band at Γ point and X point, i.e., E_Γ–E_X. Accordingly, a negative Δ indicates a direct bandgap, whereas a positive Δ indirect bandgap. Δ values of [112] SiNWs as a function of lattice constants are plotted in Fig. 4.16. It is clearly seen that Δ increases gradually from a negative to a positive value with increasing lattice constant, indicating a distinct transition from a direct to an indirect band.

The indirect bandgap of Si is known to limit its light emission, and thus applications in optoelectronics. Si nanostructures have revealed the possibility of realizing direct bandgap in Si. For example, small-diameter [110] and [111] SiNWs have been theoretically shown to be direct bandgap materials [52], and strong light emissions from them have been reported [64]. However, [112] SiNWs have been shown to have an indirect bandgap even when its diameter is smaller than 1 nm, and thus are less practical in optoelectronic applications. Thus, our finding is significant as it reveals the possibility of inducing indirect-to-direct band transition in [112] SiNWs by applying axial stress.

We further examined [112] SiNWs of different diameters, and obtained the same conclusion. [112] SiNWs of various diameters are found to exhibit similar changes in electronic energy band in response to axial stress. As axial lattice constant decreases under increasing axial compression, [112] SiNWs become a direct energy band material. However, under tension in the axial direction, SiNWs still have an indirect bandgap.

Axial compression invariably changes the internal structure of SiNWs. On axial compression, the distance between neighboring atomic layers in SiNWs along the stress direction would decrease, leading to increased overlap of atomic orbits. Thus, repulsion between atoms would develop and the total system energy increases. Meanwhile, the increased overlap of atomic orbitals would induce band splitting in conduction and valence bands. Strain-induced splitting has been found to be quite substantial in bulk Si crystal when stress is applied in [001] or [111] directions [65]. The lowest conduction state at Γ point would split into one doubly degenerate and a nondegenerate state under a strain $\delta \approx 0.02$. Our present results show similar trend, revealing the band edge of conduction band to be also non-degenerate. In addition, the amount of splitting is proportional to the stress [65]. When the splitting is large enough to push the lowest state of conduction band at Γ point below that at other points along Γ-X and the top of the valence band remains at Γ point, the original indirect bandgap would be changed to a direct one. In our calculation, the strain is large and the possibility of indirect-to-direct gap tuning is effectively increased. In particular, direct bandgap is obtained in [112] SiNWs when the strain is 0.03. In contrast, axial expansion leads to a different effect. The atomic distance between Si atoms in the neighboring layers increases, and total energy increases as well due to structure deformation. However, the electronic orbital overlap is decreased and obviously delocalization is obtained. In this case, band state splitting is not significant; consequently conduction energy band bottom is pretty flat and indirect energy band remains.

The above finding that the deformation results in the indirect–direct gap transition is of significant importance. It suggests the potential application of SiNWs in optoelectronic devices. Recent experiments found that strain exists in core–shell NWs due to a lattice mismatch between the different elements used. A low percentage of strain was found in Si/Ge core–shell NWs [66–68]. Although it is difficult to apply axial stress on a single nanowire, it is feasible to apply axial compression to a bundle of SiNWs. In fact, bundled SiNWs have been prepared by our colleagues and characterized in terms of cross-section structures [69]. Moreover, the use of nanowire bundles could enhance the light emission or light absorption. Therefore, our prediction offered important technological implications to future optoelectronics based on SiNWs.

4.3.2 Strain-Induced Band Dispersion Engineering in 2D Si Nanosheets

In the nano realm, in addition to as zero-dimensional (0D) quantum dots [70] and one-dimensional (1D) nanowires [62, 71], silicon can also appear in two-dimensional (2D) nanosheet forms [72, 73]. These low-dimensional Si nanotructures exhibit diverse electronic properties, due to quantum confinement in different dimensions, and have drawn the attention of the scientific community for extensive investigations. The past decade has witnessed a surge of research interests in the structures and properties of 2D materials in the nano realm, which has been partially boosted by the development of single-layer graphite (graphene), especially the preparation of free-standing graphene using mechanical exfoliation in 2004 [74]. Benefiting from the advancement in nanotechnology, the layered nanostructures of silicon with atomic thickness, often called nanosheet, have been successfully synthesized [72, 73, 75, 76]. Very recently, the silicon nanoribbons grown epitaxially on silver (110) and (100) surfaces paved the way for the synthesis of graphene-like silicon nanosheets [77–79]. Nanosheets bridge the gap between 1D nanomaterials and 3D macroscale bulk materials, and will thereby advance our understanding of the quantum confinement effect across the boundaries of all dimensions in materials science. Even in the same dimension, the subtle shape difference can lead to totally different properties. For example, in its 1D structure, the electronic band structure of SiNWs depends strongly on the wire diameter and growth direction [80]. Therefore, it turns out that the advances in nanoelectronics will be primarily driven by the taming of the structures of Si nanomaterials for the purpose of tailoring their electronic properties.

As compared to 1D SiNWs [63, 81–85], the investigations of strain effect on 2D Si nanosheets are relatively few, partly due to the difficulties experienced in the synthesis of free-standing Si nanosheets [73, 76, 78]. Herein, we explore the possibility of using strain to manipulate the electronic band structures of Si nanosheets, while focusing on its most widely investigated (100) and (110) facets. Our results indicate that strain can tune not only the magnitude of the bandgap, but also the band dispersion (i.e., direct or indirect bandgap), which has crucial implications for the applications of Si nanosheets in optoelectronics, luminescence, and solar cells.

The application of tensile and compressive axial strains were simulated by scaling the lattice constant along the two mutually perpendicular directions of the nanosheet, following the study of the strain effects on nanowires [63, 81–85]. The positive values of strain refer to expansion, while negative corresponds to contraction. After changing the box size, the coordinates of all atoms were allowed to relax to the equilibrium structure. The extent of applied strain considered in our computational investigations was ± 10 %. A considerably large strain studied in theoretical calculations may not be easily realizable in experiments. However, the trend predicted by first-principles calculations would facilitate a clearer understanding of the underlying physics, which in turn would provide useful pointers to experimental efforts.

The (100) Si nanosheet is stretchable along two equivalent <110> directions. Consequently, there exist two modes for applying axial strain on it: (i) symmetrical strain along x and y directions and (ii) asymmetrical strain along x or y direction. When an asymmetrical strain is applied to the Si nanosheet along x, the nanosheet will contract along the y direction. Along x direction, the ratio of changed lattice constant to strain-free, optimized lattice constant refers to the applied strain. For a fixed lattice constant along x direction, a series of total energy calculations have been performed using different lattice constants along y direction. Along y direction, the lattice constant corresponding to the energy minimum refers to the changed lattice constant, and its ratio to strain-free lattice constant is termed as the induced strain. Figure 4.17a shows the induced strain as a function of strain applied to (100) Si nanosheet. When subject to the same degree of applied strain, the nanosheet is found to be more sensitive to compressive strain than to tensile strain. For example, +5 % (−5 %) applied strain induces −0.15 % (0.3 %) strain. This implies that the (100) Si nanosheet is relatively readily compressible along the <110> direction, as compared to the other direction.

Fig. 4.17 Induced strain as a function of applied strain for **a** (100) Si nanosheet and **b** (110) Si nanosheet. The *insets* are the ball and stick models representing the atomic arrangement where the big and small spheres denote silicon atoms and hydrogen atoms, respectively. Reprinted with permission from Ref. [86]. Copyright 2011, American Chemical Society

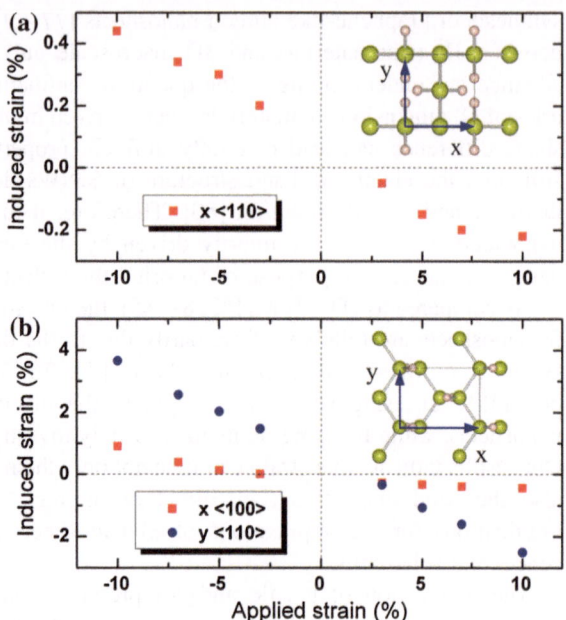

The modes of applying strain to the (110) Si nanosheet is somewhat different from that of the (100) Si nanosheet, due to the two nonequivalent directions. The structure sensitivity of the (110) Si nanosheet to strain application along <110> and <110> directions is shown in Fig. 4.17b. The applied strain along <110> (*y*) direction induces a much larger strain. In particular, 10 % compressive strain along <110> direction induces 3.7 % expansion along <110> direction, the largest induced strain. Similar to the (100) Si nanosheet, applied compressive strain induces a larger strain. For example, +5 % asymmetrical strain applied along *y* direction induces −1.1 % strain, but +5 % asymmetrical strain applied along *x* direction can produce only −0.33 % induced strain. It may be inferred from these results that the (110) Si nanosheet is more sensitive to the application of compressive stress along <110> direction.

Figure 4.18 shows both the symmetric and asymmetric strain dependences of electronic band structure of (100) Si nanosheet. As the light emission or absorption of silicon-based nanostructures depends strongly on the bandgap, we have labeled the energy states in Fig. 4.18, which crucially determine the bandgap, such as C_1 for the energy state at the conduction band edge. The energy states at Y point show a behavior similar to that at Γ point under strain, but they do not play a dominant role in bandgap transition and are therefore omitted in the following discussion. When no strain is applied to the (100) Si nanosheet, the energy differences between V_1 and V_2 ($\Delta V = V_1-V_2$) and the same between C_1 and C_2 ($\Delta C = C_2-C_1$) are 0.36 and 0.50 eV, respectively. It is noted here that a positive value of ΔV (ΔC) implies that the energy V_1 (C_2) is higher than V_2 (C_1). If both ΔV and ΔC are

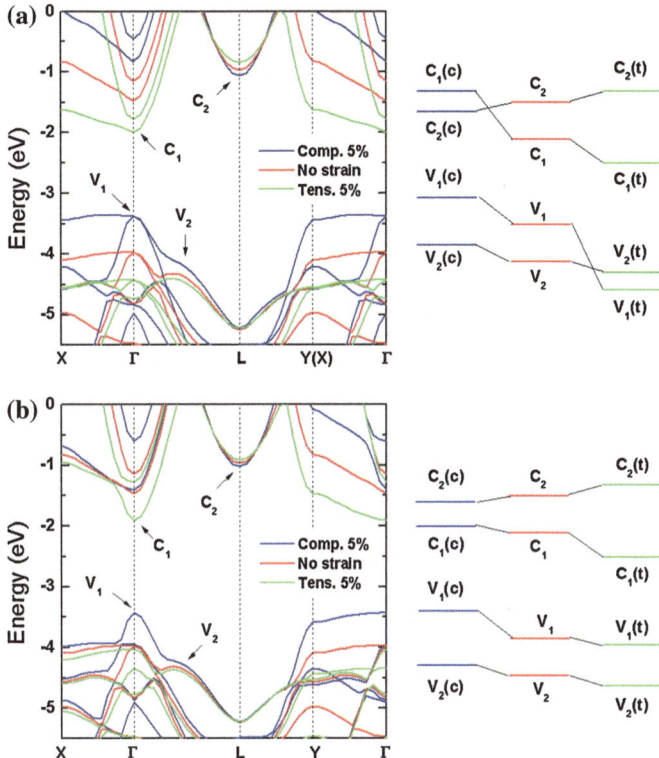

Fig. 4.18 Electronic band structure (*left panel*) and schematic description of the evolution of energy states (*right panel*) of (100) Si nanosheet under **a** symmetrical strain and **b** asymmetrical strain. The "c" and "t" in parenthesis denote compressive and tensile strains, respectively. Reprinted with permission from Ref. [86]. Copyright 2011, American Chemical Society

positive, a direct bandgap is favored for (100) Si nanosheet, showing potential for optoelectronic applications. When symmetrical compressive strain is applied to the (100) Si nanosheet; the C_1, V_1, and V_2 states rise in energy, while C_2 state drops energetically. When the compressive strain is large enough to induce a negative ΔC value, the Si nanosheet undergoes a direct-to-indirect transition. Contrary to compressive strain effect, the C_1, V_1, and V_2 states are downshifted in energy when symmetrical tensile strain is applied to the Si (100) nanosheet. Interestingly, the magnitude of downward shift of V_1 is much larger than that of V_2, yielding a negative ΔV. Based on these, it can be concluded that both symmetrical compressive and tensile strains can easily alter the electronic band dispersion of (100) Si nanosheet from a direct bandgap to an indirect bandgap.

The effect of asymmetrical strain on band structure of (100) Si nanosheet is somewhat different from that of symmetrical strain. Similar to the case of symmetrical compressive strain, the C_1, V_1, and V_2 states are shifted up in energy while the C_2 state is downshifted. However, the extent of upward shift of C_1 state is small, leading to a positive ΔC, such as 0.38 eV at 5 % compressive strain. Thus, the (100) Si nanosheet preserves the direct bandgap up to 10 % compressive strain. When asymmetrical tensile strain is applied to the Si nanosheet along <110> direction, all these involved energy states show totally different behaviors, leading to positive values for both ΔC and ΔV. Thus, within our studied range of applied strain, we observe that the application of asymmetrical compressive strain retains the direct bandgap dispersion of (100) Si nanosheet, while the tensile strain enhances the direct bandgap.

The different strain dependences of V_1, V_2, C_1, and C_2 states can be understood from decomposed charge density, as shown in Fig. 4.19. The decomposed charge density of C_1 state shows a zigzag spatial distribution, surrounding closely the Si atoms from the two mutually perpendicular directions. Interestingly, the zigzag-shaped decomposed charge density occurs parallel to the Si bonds. The decomposed charge density of V_1 state encircles the Si–Si bonds and Si–H bonds. The decomposed charge densities of C_1 and V_1 states reflect the bonding character of the (100) Si nanosheet, and are, therefore, very sensitive to structural changes. In response to the variation of the structures of Si nanosheets with the external strain, the energy of C_1 and V_1 states change rapidly. Moreover, both the decomposed charge densities of C_1 and V_1 states distribute along two directions. The symmetrical strain can induce a considerable change. The decomposed charge density of the C_2 state distributes in the interatomic space between the Si atoms, while the decomposed charge density of V_2 distributes mainly in the xy plane of the nanosheet and normal to it (i.e., the z direction). They do not relate closely to the Si–Si bonds, and consequently, the structural change of (100) Si nanosheet marginally influences the shape of decomposed charge density corresponding to C_2 and V_2.

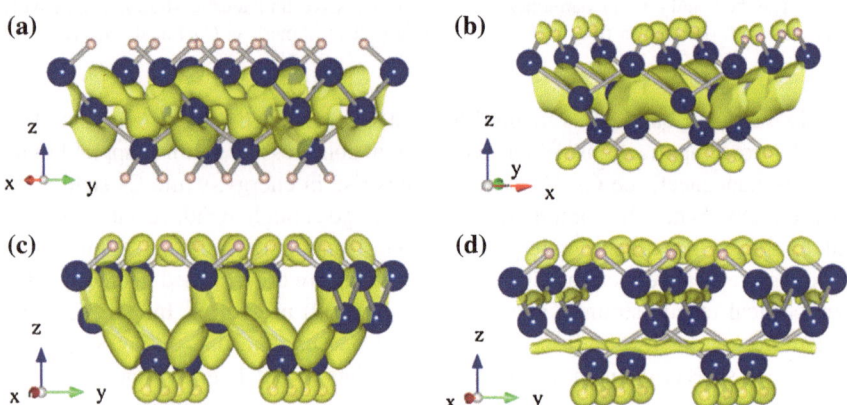

Fig. 4.19 The isosurfaces of the electronic charge density corresponding to **a** C_1, **b** C_2, **c** V_1, and **d** V_2 in (100) Si nanosheet. The isosurfaces in the figure show 35 % of the maximum electronic charge density. Reprinted with permission from Ref. [86]. Copyright 2011, American Chemical Society

When the (110) Si nanosheet is free from strain, both the valence band (V_1) maximum and conduction band (C_1) minimum locate at Γ point, showing a direct bandgap. Under strain, a state in the conduction band, labeled as C_2, participates in the direct-to-indirect bandgap transition. The schematic strain dependences of these three energy states and the electronic band structure of the (110) Si nanosheet are shown in Fig. 4.20. The behaviors of V_1 under symmetrical and asymmetrical strains are identical, as shown in the right panel of Fig. 4.20. The energy of V_1 state increases when subject to compressive strain, while it decreases under tensile strain, similar to the V_1 behavior in the (100) Si nanosheet. The C_2 energy state shows weak strain dependence, regardless of whether it is tensile or compressive strain. Thus, the other conduction band energy state, C_1 located at the Γ point, dominates the bandgap characteristic. When symmetrical compressive strain is applied to the (110) Si nanosheet, the energy of C_1 rises. The ΔC becomes negative under sufficient compressive strain. As a result, the (110) Si nanosheet experiences a direct-to-indirect bandgap transition. Contrary to the effect of compressive strain, the energy of the C_1 state decreases remarkably under symmetrical tensile strain, enlarging the direct bandgap and its characteristics.

Interestingly, the energy of C_1 state scarcely changes under asymmetrical strain along x direction. The (110) Si nanosheet not only preserves the direct bandgap characteristic, but also shows amazing properties under asymmetrical strain. Under asymmetrical strain along y direction, the electronic band structure of (110) Si nanosheet is similar to that under symmetrical strain. The compressive strain induces a negative ΔC, and the tensile strain contributes positively to ΔC, enhancing the direct bandgap.

The different strain dependences of C_1, C_2, and V_1, particularly the invariance of C_1 under asymmetrical strain along x direction, can also be explained through decomposed charge density analysis. The semi-lunar shaped decomposed charge density corresponding to the C_1 energy state encloses each Si atom from one direction. The decomposed charge density of the C_1 distributes along x direction, and a large part of it approaches the center of distorted Si hexagon when viewed from z direction, as shown in Fig. 4.21(a). Applying asymmetrical strain along x direction, the shape of decomposed charge density at C_1 does not change significantly. However, the asymmetrical strain along y direction will affect the distribution of decomposed charge density at C_1 qualitatively. Therefore, the energy of the C_1 state barely changes under asymmetrical strain along x, as shown in Fig. 4.20b. The decomposed charge density of V_1 encloses the Si–Si bond in the xy plane. A careful inspection of the shape of decomposed charge density corresponding to V_1 reveals that its distribution projected along the x direction is larger than that projected along y, enhancing the sensitivity of the energy of V_1 to asymmetrical strain along y. With the same magnitude of strain, the increment of V_1 under asymmetrical strain along y is larger than that under asymmetrical strain along x, as shown in the Figs. 4.20b, c. Similar to the distribution of the C_2 in the (100) Si nanosheet, the decomposed charge density of the C_2 in the (110) Si nanosheet distributes itself in the interatomic space of Si atoms, and contributes weakly to Si–Si bonds. Thus, the energy of the C_2 state shows weak strain

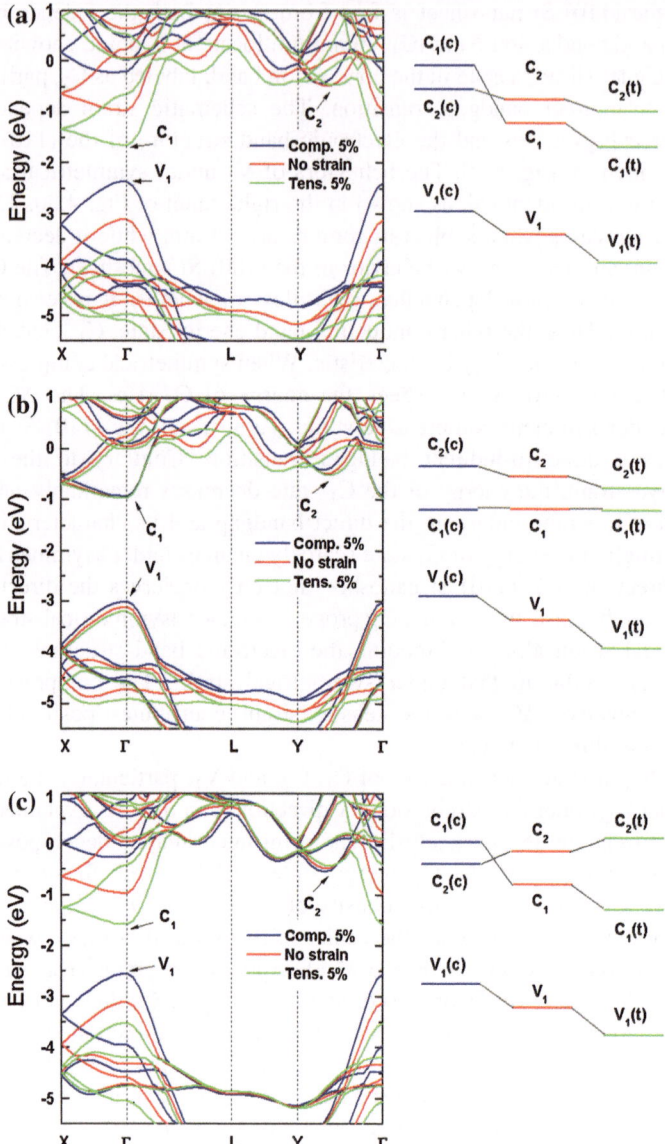

Fig. 4.20 Electronic band structure (*left panel*) and schematic representation of the evolution of energy states (*right panel*) of (110) Si nanosheet under **a** symmetrical strain and **b** asymmetrical strain along *x* direction, and **c** asymmetrical strain along *y* direction. The "c" and "t" in parenthesis denote compressive and tensile strains, respectively. Reprinted with permission from Ref. [86]. Copyright 2011, American Chemical Society

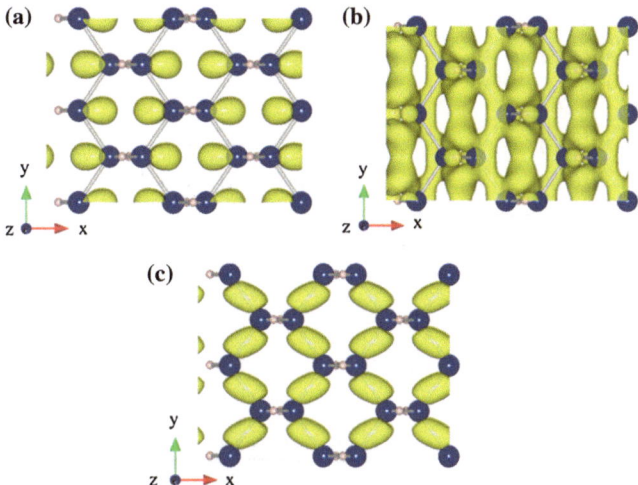

Fig. 4.21 The isosurfaces of electronic charge density at **a** C_1, **b** C_2, and **c** V_1 of the (110) Si nanosheet. The isosurfaces are 35 % of the maximum density. Reprinted with permission from Ref. [86]. Copyright 2011, American Chemical Society

dependence. The variation of the bandgap of Si nanosheet with strain is shown in Fig. 4.22. While DFT with GGA functional is known to underestimate the bandgap, more accurate quantum Monte Carlo calculations [33] and GW approximation [52, 87] have shown that DFT reproduces the general trends of bandgap variation. When free from strain, the (100) Si nanosheet tends to have a larger indirect bandgap than (110) Si nanosheet due to the larger quantum confinement. This trend holds good for symmetrically strained nanosheets. Therefore, unstrained (100) Si nanosheets have a stronger potential for solar cell applications than the (110) nanosheets because of the larger indirect bandgap of the former.

Figure 4.22 shows clearly the strain tunability of bandgap of Si nanosheets. To obtain a smaller bandgap, a larger strain (at least 5 %) needs to be applied to the (100) Si nanosheet, as shown in Fig. 4.22a. By carefully controlling the magnitude of applied symmetrical compressive strain to the (100) Si nanosheet, a blueshift followed by a redshift of the spectrum can be induced in photoluminescence measurements. This owes its origin to the rise in energy of C_1 state. To have a direct bandgap in optical and luminescence applications, asymmetrical strain along x direction, as shown in Fig. 4.22a by filled squares, will have to be applied.

The variation of the bandgap of (110) Si nanosheet with strain is shown in Fig. 4.22b. Interestingly, the magnitude of the direct bandgap of the (110) Si nanosheets increases monotonically and linearly with asymmetrical strain along x direction. Based on this behavior, one can get a larger bandgap by applying tensile strain and a smaller gap by compressive strain. The unique bandgap dependence accrues mainly from the C_1 energy state, whose energy remains almost nearly constant under asymmetrical strain along x. The smaller bandgaps of

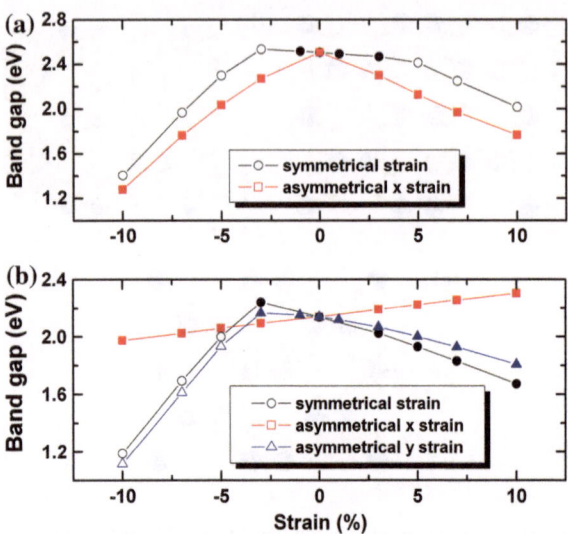

Fig. 4.22 Bandgap versus strain for **a** the (100) Si nanosheet and **b** the (110) Si nanosheet. *Filled and unfilled symbols* denote the direct and indirect bandgaps, respectively. Reprinted with permission from Ref. [86]. Copyright 2011, American Chemical Society

direct type can also be obtained by applying symmetrical tensile strain and asymmetrical *y* tensile strain. Similar to the (100) Si nanosheet, a blueshift followed by a redshift will be observed in the spectrum, when symmetrical strain and asymmetrical compressive strain are consecutively applied along *y* direction to the (110) Si nanosheet. This underscores the importance of strain engineering in Si nanosheets for applications in optoelectronics and luminescence. Furthermore, strain engineering of Si nanosheets can be utilized more conveniently and flexibly in its solar cell applications. The external strain can be realized experimentally by controllable deposition or epitaxial growth methods. For instance, Kim et al. [73] have observed strain in their free-standing Si nanosheets and strain-induced variation in optical properties.

In our work of <112> SiNWs [63], we have found the axial compression or expansion to be effectual in altering the electronic bands near the Fermi energy level and eventually in inducing an indirect-to-direct bandgap transition under a certain degree of strain, exhibiting the potential of SiNWs for applications in optoelectronics and luminescence. In comparison to SiNWs, the 2D Si nanosheets offer more accessible ways for applying strain to it. By carefully controlling the magnitude and direction of strain application, one can obtain different band dispersions (e.g., direct or indirect bandgap) and bandgaps of different magnitudes. Therefore, strain application can be an effective way for engineering the electronic band structure of silicon nanostructures.

References

1. Cullis AG, Canham LT, Calcott P (1997) J Appl Phys 82:909
2. Kim BH, Cho CH, Kim TW, Park NM, Sung GY (2005) Appl Phys Lett 86:091908
3. Wolkin MV, Jorne J, Fauchet PM (1999) Phys Rev Lett 82:197
4. Kanemitsu Y (1994) Phys Rev B 49:16845
5. Tilley RD, Warner JH, Yamamoto K, Matsui I, Fujimori H (2005) Chem Commun (Cambridge) 14:1836
6. Wilcoxon JP, Samara GA, Provencio PN (1999) Phys Rev B 60:2704
7. Garrido M, Lopez O, Gonzalez A, Perez-Rodriguez JR (2000) Morante and C. Bonafos. Appl Phys Lett 77:3143
8. Holmes JD, Ziegler KJ, Doty RC, Pell LE, Johnston KP, Korgel BA (2001) J Am Chem Soc 123:3743
9. Belomoin G, Therrien J, Nayfeh M (2000) Appl Phys Lett 77:779
10. Wang X, Zhang RQ, Niehaus TA, Frauenheim Th, Lee ST (2007) J Phys Chem C 111:12588
11. Li QS, Zhang RQ, Lee ST, Niehaus TA, Frauenheim Th (2008) J Chem Phys 128:244714
12. Delerue C, Allan G, Lannoo M (1993) Phys Rev B 48:11024
13. Wang LW, Zunger A (1994) J Phys Chem C 98:2158
14. Yu DK, Zhang RQ, Lee ST (2002) J Appl Phys 92:7453
15. Delley B, Steigmeier EF (1993) Phys Rev B 47:1397
16. Ren SY, Dow JD (1992) Phys Rev B 45:6492
17. Hirao M, Uda T (1994) Surf Sci 306:87
18. Delley B, Steigmeier EF (1995) Appl Phys Lett 67:2370
19. Wang X, Zhang RQ, Lee ST, Niehaus TA, Frauenheim Th (2007) Appl Phys Lett 90:123116
20. Porezag D, Frauenheim TH, Köhler Th, Seifert G, Kaschner R (1995) Phys Rev B 51:12947
21. Elstner M, Porezag D, Jungnickel G, Elsner J, Haugk M, Frauenheim Th, Suhai S, Seifert G (1998) Phys Rev B 58:7260
22. Chelikowsky JR, Kronik L, Vasiliev I (2003) J Phys: Condens Matter 15:R1517
23. Prendergast D, Grossman JC, Williamson AJ (2004) J Am Chem Soc 126:13827
24. Puzder A, Williamson AJ, Grossman JC, Galli G (2003) J Am Chem Soc 125:2786
25. Hirao M, Uda T (1994) Int J Quantum Chem 52:1113
26. Porter AR, Towler MD, Needs R (2001) J Phys Rev B 64:035320
27. Weissker H-Ch, Furthmüller J, Bechstedt F (2002) Phys Rev B 65:155328
28. Onida G, Reining L, Rubio A (2002) Rev Mod Phys 74:601
29. Hahn PH, Schmidt WG, Bechstedt F (2005) Phys Rev B 72:245425
30. Benedict LX, Puzder A, Williamson AJ, Grossman JC, Galli G, Klepeis JE, Raty J-Y, Pankratov O (2003) Phys Rev B 68:085310
31. Vasiliev I (2003) Phys Status Solidi B 239:19
32. Vasiliev I, Ogut S, Chelikowsky JR (2001) Phys Rev Lett 86:1813
33. Williamson AJ, Grossman JC, Hood RQ, Puzder A, Galli G (2002) Phys Rev Lett 89:196803
34. Franceschetti A, Pantelides ST (2003) Phys Rev B 68:033313
35. Luppi E, Degoli E, Cantele G, Ossicini S, Magri R, Ninno D, Bisi O, Pulci O, Onida G, Gatti M, Incze A, Sole ED (2005) Opt Mater 27:1008
36. Niehaus TA, Suhai S, Sala FD, Lugli P, Elstner M, Seifert G, Frauenheim Th (2001) Phys Rev B 63:085108
37. Fehér F (1977) Molekülspektroskopische Untersuchungen auf dem Gebiet der Silane und der Heterocyclischen Sufane, Forschungsbericht des Landes Nordrhein-Westfalen. Westdeutscher, Köln
38. Garoufalis GS, Zdetsis AD (2001) Phys Rev Lett 87:276402
39. Zdetsis AD (2006) Rev Adv Mater Sci 11:56
40. Becke AD (1993) J Chem Phys 98:5648
41. Lee C, Yang W, Parr RG (1988) Phys Rev B 37:785
42. Puerto, MLD, Jain M, Chelikowski JR (2010) Phys Rev B 81:035309

43. Puzder A, Williamson AJ, Reboredo FA, Galli G (2003) Phys Rev Lett 91:157405
44. Allan G, Delerue C, Lannoo M (1996) Phys Rev Lett 76:2961
45. Sun J, Song J, Zhao Y, Liang WZ (2007) J Chem Phys 127:234107
46. Draeger EW, Grossman JC, Williamson AJ, Galli G (2004) J Chem Phys 120:10807
47. Hamel S, Williamson AJ, Wilson HF, Gygi F, Galli G, Ratner E, Wack D (2008) Appl Phys Lett 92:043115
48. Wang X, Zhang RQ, Lee ST, Frauenheim Th, Niehaus TA (2008) Appl Phys Lett 93:243120
49. Nayfeh MH, Rigakis N, Yamani Z (1997) Phys Rev B 56:2079
50. Wang X, Zhang RQ, Lee ST, Frauenheim Th, Niehaus TA (2009) Appl Phys Lett 94:029902
51. Wang Y, Zhang RQ, Frauenheim Th, Niehaus TA (2009) J Phys Chem C 113:12935
52. Zhao XY, Wei CM, Yang L, Chou MY (2004) Phys Rev Lett 92:236805
53. Read AJ, Needs RJ, Nash KJ, Canham LT, Calcott PDJ, Qteish A (1992) Phys Rev Lett 69:1232
54. Soler JM, Artacho E, Gale JD, García A, Junquera J, Ordejón P, Sánchez-Portal D (2002) J Phys: Condens Matter 14:2745
55. Lu AJ, Zhang RQ, Lee ST (2008) Nanotechnology 19:035708
56. Lu AJ, Zhang RQ, Lee ST (2008) Appl Phys Lett 92:203109
57. Persson MP, Xu HQ (2004) Nano Lett 4:2409
58. Zhang RQ, Lifshitz Y, Ma DDD, Zhao YL, Frauenheim Th, Lee ST, Tong SY (2005) J Chem Phys 123:144703
59. Haugerud BM, Bosworth LA, Belford RE (2003) J Appl Phys 94:4102–4107
60. Lyons DM, Ryan KM, Morris MA, Holmes JD (2002) Nano Lett 2:811–816
61. Audoit G, Mhuircheartaigh ÉN, Lipson SM, Morris MA, Blau WJ, Holmes JD (2005) J Mater Chem 15:4809–4815
62. Morales AM, Lieber CM (1998) Science 279:208–211
63. Lu AJ, Zhang RQ, Lee ST (2007) Appl Phys Lett 91:263107
64. Huo J, Solanki R, Freeouf JL, Carruthers JR (2004) Nanotechnology 15:1848
65. Tsay YF, Bendow B (1977) Phys Rev B 16:2663
66. Goldthorpe IA, Marshall AF, McIntyre PC (2008) Nano Lett 8:4081–4086
67. Goldthorpe IA, Marshall AF, McIntyre PC (2009) Nano Lett 9:3715–3719
68. Peng X, Logan P (2010) Appl Phys Lett 96:143119
69. Li CP, Lee CS, Ma XL, Wang N, Zhang RQ, Lee ST (2003) Adv Mater 15:607
70. Wolkin MV, Jorne J, Fauchet PM, Allan G, Delerue C (1999) Phys Rev Lett 82:197–200
71. Zhang YF, Tang YH, Wang N, Yu DP, Lee CS, Bello I, Lee ST (1998) Appl Phys Lett 72:1835
72. Nakano H, Mitsuoka T, Harada M, Horibuchi K, Nozaki H, Takahashi N, Nonaka T, Seno Y, Nakamura H (2006) Angew Chem Int Edit 45:6303–6306
73. Kim U, Kim I, Park Y, Lee K-Y, Yim S-Y, Park J-G, Ahn H-G, Park S-H, Choi H-J (2011) ACS Nano 5:2176–2181
74. Novoselov KS, Geim AK, Morozov SV, Jiang D, Zhang Y, Dubonos SV, Grigorieva IV, Firsov AA (2004) Science 306:666–669
75. Okamoto H, Kumai Y, Sugiyama Y, Mitsuoka T, Nakanishi K, Ohta T, Nozaki H, Yamaguchi S, Shirai S, Nakano H (2010) J Am Chem Soc 132:2710–2718
76. Sugiyama Y, Okamoto H, Mitsuoka T, Morikawa T, Nakanishi K, Ohta T, Nakano H (2010) J Am Chem Soc 132:5946–5947
77. Kara A, Léandri C, Dávila M, De Padova P, Ealet B, Oughaddou H, Aufray B, Le Lay G (2009) J Supercond Nov Magn 22:259–263
78. Le Lay G, Aufray B, Léandri C, Oughaddou H, Biberian JP, De Padova P, Dávila ME, Ealet B, Kara A (2009) Appl Surf Sci 256:524–529
79. Aufray B, Kara A, Vizzini S, Oughaddou H, Leandri C, Ealet B, Le Lay G (2010) Appl Phys Lett 96:183102
80. Zhang RQ, Lifshitz Y, Ma DDD, Zhao YL, Frauenheim T, Lee ST, Tong SY (2005) J Chem Phys 123:144703
81. Hong K-H, Kim J, Lee S-H, Shin JK (2008) Nano Lett 8:1335–1340

82. Huang L, Lu N, Yan J-A, Chou MY, Wang C-Z, Ho K-M (2008) J Phys Chem C 112:15680–15683
83. Leu PW, Svizhenko A, Cho K (2008) Phys Rev B 77:235305
84. Logan P, Peng X (2009) Phys Rev B 80:115322
85. Peng X, Alizadeh A, Kumar SK, Nayak SK (2009) Int J Appl Mech 1:483–499
86. Zhang C, De Sarkar A, Zhang RQ (2011) J Phys Chem C 115:23682
87. Lu AJ, Zhang RQ (2008) Solid State Commun 145:275–278

Chapter 5
Summary and Remarks

To promote silicon-based nanoscience and nanotechnology, we have performed systematic computational studies on various possible silicon nanostructures, including 0D silicon quantum dots, 1D silicon nanowires, and 2D silicon sheet, for their growth mechanism, structural stability, chemical stability, excited state property, and energy band structure engineering. Specifically, we elucidated the interesting oxide assisted growth mechanism by calculating a huge number of gas-phase silicon oxide clusters and obtaining their reactivity as a function of silicon-to-oxygen ratio. By revealing a high chemical reactivity of the silicon suboxide clusters and the formation of tetrahedral core when they grow larger, we successfully obtained the nucleation and growth mechanism of silicon nanostructures from silicon oxide. In the mechanism, gas-phase silicon suboxide clusters can bond to each other through Si–Si bond formation driven by their high chemical reactivities. The so-grown clusters can develop to form tetrahedral silicon cores at high temperature. Due to the crystallographic dependent oxygen diffusion in the cluster, the chemical reactive site can be maintained at the cluster surface in a certain direction, which can absorb additional gas phase silicon oxide clusters, leading to the continuous growth in a certain direction such as <110> or <112>. Silicon nanostructures without surface saturation are high unstable and can undergo amorphization. They may form a huge number of possible morphologies include nanowires and nanotubes but with highly unstable structures, deviated largely from tetrahedral configurations. Surface saturation using H, F, Cl, N, and O, etc., can significantly improve the energetic stability of the tetrahedral silicon nanostructures, clean up the gap states due to surface dangling bonds, and enable the silicon nanostructure presenting quantum confinement effect and light emitting. In particular, surface hydrogenation is the most simple and an efficient way of surface saturation of silicon nanostructures, which could be done using conventional chemical etching techniques, for either silicon wafers or silicon nanostructures. The leaving of H rather than F on the silicon surface in the etching is due to the polarization and weakening of Si–Si bond near the surface saturated with F because the fluoridation promotes the chemical reaction further with the HF molecules in the solution. The surface-hydrogenated silicon nanostructures can be thermally very stable and also chemically stable according to our studies of chemical reactions of silicon nanostructures containing surface hydrides, with a water molecule. There is a clear-size dependent chemical stability of silicon nanostructures, according to which the smaller the size of the nanostructure is the higher the chemical stability, revealing the possibility to fabricate stable nanoscale devices based on silicon nanostructures, removing the difficulty in maintaining good chemical stability of microscale devices.

However, the hydrogenated silicon nanostructures suffer a stability problem in excited state due to the localization of the excited electron, which can significantly elongate some Si–Si bonds. This excited state instability can be reduced once the nanostructure size reaches 1.5 nm due to the strong restriction coming from numerous outer atoms that form a rigid network. Therefore, there is only a strong Stokes shift of silicon nanostructures with a size smaller than 1.5 nm. In general, the optical absorption gap of silicon nanostructures decreases with the size increase, but their emission gaps deviate significantly from such a trend at small size due to the Stokes shift. When the silicon nanostructures grow into wires in certain directions, the excited electron becomes delocalized and thus the induced instability gets reduced. To achieve good optical transition of electron for opto-electronic applications, direct energy band structures of silicon nanowires are needed. The silicon nanowires can show both direct and indirect energy band structures dependent on the orientations, sizes, and enclosing facets. In particular, <112> silicon nanowires present unique electronic structures and tunability, in which a wide (110) facet can promote the generation of direct band structures. The band structures can be also tuned by applying strain or stress. For the <112> silicon nanowires, compression is very effective in inducing direct band structures. When the dimension scale is up to two, the possibility of engineering the band structures using straining is significantly increased. Asymmetrical strain can cause a direct-to-indirect transition in (100) Si nanosheet, while symmetrical strain retains its direct bandgap characteristic. Under asymmetrical strain along <100> direction, the direct bandgap of (110) Si nanosheet exhibits unique characteristics, with the direct bandgap varying linearly with strain. Similar bandgap variation behaviors are also true for (110) Si nanosheets subject to symmetrical and asymmetrical strain along <110> direction. The various strain dependences are attributed to the changes in the nanostructure's charge density. Our findings highlight the immense potential of various Si nanostructures for applications in optoelectronics, sensing, luminescence, and solar cells.